THE NEW NATURALIST

A SURVEY OF BRITISH NATURAL HISTORY

THE SEA SHORE

The aim of this series is to interest the general reader in the wild life of
Britain by recapturing the inquiring spirit of the old naturalists. The
Editors believe that the natural pride of the British public in the native
fauna and flora, to which must be added concern for their conservation,
is best fostered by maintaining a high standard of accuracy combined
with clarity of exposition in presenting the results of modern scientific
research. The plants and animals are described in relation to their homes
and habitats and are portrayed in the full beauty of their natural colours,
by the latest methods of colour photography and reproduction.

THE NEW NATURALIST

THE SEA SHORE

by

C. M. YONGE

D.Sc. F.R.S.E. F.R.S.

REGIUS PROFESSOR OF ZOOLOGY IN THE
UNIVERSITY OF GLASGOW
CHAIRMAN OF COUNCIL, SCOTTISH MARINE
BIOLOGICAL ASSOCIATION

WITH 61 COLOUR PHOTOGRAPHS
BY D. P. WILSON AND OTHERS
62 BLACK-AND-WHITE PHOTOGRAPHS
AND 88 TEXT FIGURES

COLLINS   ST. JAMES'S PLACE   LONDON

First published in 1949 by
Collins 14 St. James's Place London
Produced in conjunction with Adprint
and printed in Great Britain
by Collins Clear-Type Press
London and Glasgow

IN MEMORY

M. J. Y.

WHO WILL WALK
ON NO MORE SHORES
WITH ME

ERRATA

Colour Plates

9a   For *Tealiaf* read *Tealia*

25a   Caption should read as for 38a

25b   Caption should read as for 25a

38a   Caption should read as for 25b

CONTENTS

COLOUR PLATES

It should be noted that throughout this book Plate numbers in arabic figures refer to Colour Plates, while roman numerals are used for Black-and-White Plates.

PLATES IN BLACK AND WHITE

All black-and-white photographs were taken by D. P. Wilson unless otherwise stated

TEXT FIGURES

xiv

EDITORS' PREFACE

OF all the habitats in Britain in which communities of animals and plants are found, none is richer than the sea shore. This narrow strip, extending for thousands of miles (though covering a relatively small area), is occupied by an astounding variety of life. One of the main reasons for this is that the habitat is divided horizontally into zones which depend primarily upon the frequency and duration of submergence; the effect of the tides largely determines the natural riches of the sea shore. Further, rock, sand and mud provide radically different environments and the sea-shore communities associated with these differ widely from one another.

It is at the sea shore, along the edge of Britain, that the ordinary observer gets his only opportunity of encountering animals and plants belonging to many classes and groups which are almost or entirely confined to the sea. Such groups include the Echinoderms, which are entirely restricted to salt water and contain the sea-urchins, starfish, brittle stars and sea-cucumbers; the Tunicates, also entirely marine; the Polyzoa—nearly all marine, as are the jellyfish. Corals and sea-anemones are never found in fresh water. The octopuses, cuttlefish and squids, the most highly organised of the molluscs, are entirely marine. Among plants the seaweeds are almost restricted to the sea shore and the shallow sea.

The sea shore thus forms a natural laboratory for the student, because here he can obtain the best possible introduction to a general knowledge of animals and plants. The interests of C. M. Yonge, who now occupies the Regius Chair of Zoology in the University of Glasgow, are entirely marine. He began his research career on the staff of the marine biological laboratory at Plymouth, has worked at marine stations all over the world, including one established by the expedition he led to the Great Barrier Reef of Australia in 1928-29, and is now chairman of the committee that conducts the marine biological laboratory at Millport on the Clyde. That Professor Yonge is a master of exposition readers of this book will be able to judge; that he has performed a formidable amount of research they will also discover for themselves.

EDITORS' PREFACE

We consider that *The Sea Shore* is one of the most important books we have had the privilege of publishing in this series. We feel this not only because of Professor Yonge's felicitous treatment of his complicated subject, but because of the subject's own value. The end to which the NEW NATURALIST series of books is devoted is the exposition of British Natural History from the point of view of ecology, and with an accent on the communities of living things. In the study of shore life the ways in which animals and plants are members of communities, and their adaptations to their environments, are specially complex and interesting. As an introduction to general ecology there is nothing better than a study of the basic subjects of geology and shore life ; and we are glad that we have been able to publish books on both of these subjects early in our series.

THE EDITORS

CHAPTER I

INTRODUCTION

"As natural history has, within the last half century, occupied the attention and pens of the ablest philosophers of the more enlightened parts of the globe, there needs no apology for the following sheets ; since the days of darkness are now past, when the researches of the naturalist were considered as trivial and uninteresting."

GEORGE MONTAGU : *Testacea Britannica.* 1803

M A N Y books have been written about the natural history of British shores. There are very many more dealing with the structure, habits and classification of the numerous groups of animals and plants which are found upon them. The shore, here regarded as the region between extreme high and extreme low spring tides,[1] may at first sight appear but a small area to have received so much attention at the hands of naturalists. There are somewhat over 600,000 acres of shore around the British Isles, but this is little compared with the land mass of these islands and almost negligible in comparison with the vast expanse of the surrounding seas. But this narrow strip holds an interest altogether out of proportion to its extent. This is due in the first place to the unique richness of its population ; in the second place to the fact that here alone we can see and handle the inhabitants of the sea ; in the third place to the diversity and beauty of the adaptations which have fitted its inhabitants for life on the shore.

The sea shore is the meeting place of sea and land. It is for that reason the most fascinating and the most complex of all the environments of life. It has long been recognised as the best training ground for a zoologist because upon it are to be found members of almost all of the invertebrate groups of animals and with them shore fishes and even, on occasion, marine mammals such as stranded whales or breeding seals. On the shore exposed by the ebbing tide hosts of sea-birds descend to feed. Rocky shores are clothed with rich carpets of green, brown and red seaweeds. All of these animals and plants dwell

---

[1] Throughout this book the shore is regarded as the region bounded by extreme tidal movements.

       B

in a region which fluctuates in extent from day to day as tides wax and wane from springs to neaps and back again, so that what is shore one day may be either dry land or continuously submerged the next. Within the narrow confines of what is always covered and uncovered daily by the sea, conditions change from hour to hour as stranded life is re-submerged under the advancing tide or what was covered is left exposed by the receding waves. Changes no less great than those which slowly overtake land animals as season gives way to season are thus experienced twice daily by the creatures of the shore.

Here is not one environment but many ; the differences between the conditions for life on shores of rock, of sand and of mud, even between various areas within these, are as great as the differences between woods, moorlands and meadows on land. These differences are reflected in the widely varying nature of the inhabitants of each type of shore. On a single stretch of shore, moreover, the differences between the population at low water level and that at high water level are comparable with those between the inhabitants of the plain and of the mountain top. The zonation of life which spreads over thousands of ascending feet on land is, on a rocky shore, telescoped within a few score yards.

Anyone who would tell of the life on British shores is thus faced with a vast assemblage of the most diverse forms of life and also with a series of widely different environments for each of which the inhabitants are differently adapted. Some idea must be conveyed to the reader of the inter-relationships of the shore population, both animal and plant, by describing what is known about the organisation of this complex society. The inhabitants of the shore are no longer to be regarded just as the " specimens " which were so ardently collected by the Victorian naturalist. They are living things maintaining themselves in often precarious and transitory equilibrium with the manifold forces of the physical and biological environment. Each one unconsciously strives for food and protection and for the ultimate goal of all life : reproduction and the perpetuation of its kind.

It is already, I hope, becoming clear that the major problem facing the author of this book is one of selection. The subject-matter of this volume demands very different treatment from one dealing with a single group of land animals, such as birds or mammals, or with life in a single type of environment, highly complex though this may be, such as a pinewood or a meadow. There exists, it is pleasant to know, a large and growing number of people who have an excellent general knowledge of British birds, mammals, freshwater fishes and certain groups of insects. But, outside the insects, few invertebrate groups interest the amateur naturalist in these days ; conchologists and carcinologists are rare while still fewer have more than a nodding acquaintance with segmented worms, with starfish and sea-urchins, or with the hydroid and medusoid coelenterates. The merest handful, apart from professional zoologists, can claim even to identify as such a nemertine worm, a polyzoan colony or a sea-squirt. Yet these are among the commonest animals on our shores. Again, the environmental factors on land, consisting essentially of the daily and seasonal changes in temperature, light and humidity, are experienced by each one of us and we live in the same medium of air as the other inhabitants of the land. It is by no means so easy for us to realise the significance of rapidly changing salinity and of alternate submergence and exposure, or the effect of masses of tempest-driven water pouring over a sandy beach or dashing against a rocky shore. But to the animals of the shore these represent the daily hazards of life and each must be adapted to surmount or circumvent the dangers and difficulties they represent.

It must be realised from the outset that, although exposed in some measure to the air for varying periods daily, the animals and plants of the shore are, with but few exceptions, marine organisms. A few it is true, following a path blazed by pioneers millions of years ago, are in process of adapting themselves to life on land. The shore has always been one of the training grounds for those unconsciously seeking to graduate as inhabitants of the land. There is a region around high tide mark where there is a mingling of the inhabitants of the land with those of the sea. Some insects, essentially terrestrial animals, live only along the fringes of the shore, a smaller number descending into the tidal region. There is a large and characteristic community of flowering plants, with some algae and lichens, which grow on sand-dunes or on cliffs where they are constantly exposed to the salt spray driven inland by the wind. But all of these, like the birds which feed on the shore or nest on the cliffs, are inhabitants of the land and, apart from incidental mention, lie outside the scope of this book. Another mingling comes in the estuaries. While the true inhabitants of the sea and of fresh waters are sharply distinguished, the mingled, brackish waters of estuaries contain a population derived from both sources and in so far as these are migrants from the sea, they must be mentioned.

Much is to be learnt from them about the problems of passage from sea to freshwater, a route taken in the distant past by our own remote ancestors in their eventual passage from sea to land.

The distinction between the inhabitants of shore and sea is the most difficult of all. There is a large group of truly shore, or littoral, animals and plants, for instance certain acorn-barnacles and sea-anemones, with marine snails such as dog-whelks and limpets, and various worms and crustaceans, together with the intertidal seaweeds. But many shallow water animals may be found on the shore, especially if there are standing pools, shelter under rocks and weed, or sand into which they can burrow. Other animals, such as prawns and edible crabs, come inshore in the summer and retreat into deeper water with the onset of winter. Exceptionally low spring tides may also briefly uncover animals and plants which cannot be considered as typical members of the fauna or flora of the shore. Many of these shallow water animals are common on the shore and must be described. It must be remembered that there is no sharp distinction between the population above extreme low water level and in the shallow waters off-shore.

Like its fellows in this series, this book aims to interest a wide public. The student and intending specialist can seek knowledge in strictly scientific books and memoirs. The general public is perhaps less fortunate. I have gained so much pleasure from studying the shores and shallow seas around these islands and also along the coasts of western and southern Europe, of America and of tropical Australia, that I welcome this opportunity to pass on to others some measure of my own enjoyment. I shall be well satisfied if this book adds something to the interest so many experience when strolling along a sandy beach or scrambling over seaweed-covered rocks and viewing the life uncovered by the tide. My main endeavour has been to stress the broad problems of shore life and the unconscious ingenuity, which we call adaptation, by which shore animals have overcome the hazards of their ever-changing environment. I have made no attempt to catalogue, still less to describe, the multitude of species of animals and plants which industrious search will reveal between tide marks.

PLATE I
TYPICAL EXPOSED ROCKY SHORE, Trevone, North Cornwall, showing Limpets, *Patella vulgata* and Acorn-barnacles, *Chthamalus stellatus*, with stunted Bladder-wrack, *Fucus vesiculosus*

PLATE I

PLATE 2

Photographs by D. P. Wilson

Guidance for those who seek further knowledge is provided by frequent references in the text to books dealing with particular groups of animals, while further bibliographical information is given in the Appendix, but I have purposely refrained from breaking the flow of the narrative by more than occasional mention of the authors originally responsible for the observations quoted.

Inevitably many animals and plants must be mentioned but it is the really common and well-known animals, the limpets and peri-winkles, the mussels and cockles, the barnacles and crabs, the jellyfish and sea-anemones, the starfish and sea-urchins, the lug-worms and ragworms, which dominate the stage. Their distribution and their habits will illustrate the major themes.

Such a treatment while, it is hoped, making more general appeal, does also reflect the changed attitude of naturalists in their approach to the problems of shore life. While the discovery of a new species may still occasionally reward the diligent, this initial phase of description is now largely completed. The interest of those who study the life of the shore has turned from classification and structure to function and behaviour, and from a consideration of individual animals to that of the community of which they are members and on which they are dependent.

The history of the growth of knowledge about the natural history of our shores and seas has never been adequately told. This is not the place to repair this omission but at least the outlines of the story may be related. The way will then be opened to a brief account of the types of living things found and described by the early naturalists from whose labours there gradually arose the massive edifice of the modern science of oceanography, of which the study of shore life forms a part. And for that reason we must for a short space pass beyond the confines of the shore to consider the basic factors controlling the fertility of the seas and so fill in the backcloth to the actions of our actors on the sea shore.

PLATE 2

a. ENTEROMORPHA POOL at extreme high water level of spring tides

b. CORNISH ROCK POOL, mid-tide level, with a rich flora especially of *Bifurcaria tuberculata* but including *Fucus serratus*, *Enteromorpha* and *Corallina officinalis*

Like all life, that of the shore has beauty and, as always, that beauty is greatest in its natural setting. An expanded sea-anemone or a swimming prawn have grace and beauty when viewed through the glass sides of an aquarium tank but this beauty is enhanced in a rock pool through parted vegetation of green and red weeds. There are few more beautiful sights than a rock pool or the under-side of an overhanging boulder covered with encrusting growths of coloured sponges and compound sea-squirts. This beauty, which adds so greatly to the joy of the shore naturalist, is not easily conveyed in words. Where these fail the photography of my friend, Mr. D. P. Wilson, has brilliantly succeeded. It is a pleasure to conclude this introductory chapter by a record of my gratitude to him for the beauty of his photographs which largely illustrate this book. I should also like to express my appreciation of the skill of Miss Alison Birch, who was responsible for most of the drawings. To others who have assisted with photographs and with figures, including in the latter Mrs. E. W. Sexton and Mr. F. S. Russell, F.R.S., of the Marine Biological Laboratory, Plymouth, and also my daughter, Elspeth, I render my thanks. Final acknowledgments are due to Dr. H. Höglund of Lysekil, Sweden, whose photographs of prawns are both a major addition to the illustrations of this book and an evidence of the fellowship which unites those who study the life of the shore and of the sea in all lands.

CHAPTER 2

THE DISCOVERY OF THE SEA SHORE

"And now, gentle reader, let us hasten to the beach : the tide is near its ebb, and yonder rocks, baring their shoulders to the sunshine, seem to rest themselves in grim repose.

" This is the time for work. Come, boy ! the fishing basket and the muslin landing-net—a hammer and an iron chisel. Mind, too, you don't forget the large glass jar with handles made of rope, wherein to put what specimens we find."

THOMAS RYMER JONES : *The Aquarian Naturalist.* 1858

T H E G R E E K S, founders of the sister sciences of zoology and botany, dwelt beside tideless shores. Their astonishment would have been as great as was that of Hannibal and his Carthaginians at Cadiz when faced with the great rise and fall of tides in the open ocean. The Romans found amusement in the collection of shells washed up on the shore and in the construction of fish ponds (piscinae) where they kept mullet, bass, turbot and the handsome and savage roman or moray eel. But they had not the scientific outlook, and the sound foundations built by the Greeks, notably by Aristotle, were overgrown by weeds of credulity in Roman and still more in medieval times when every traveller's tale was accepted as truth.

For two centuries after the Renaissance the early naturalists struggled to clear away this growth of superstition and to resume the building of a sound edifice of natural knowledge. Then, in the eighteenth century, came the great systematist, Carl Linnaeus. He brought order into the chaos of the animal and plant kingdoms by establishing in common use the modern methods of naming and describing species. By so doing he gave an immense impetus to the collection and scientific classification of new members of the fauna and flora of the world. No small part of the wealth and leisure of cultured men in western Europe was so employed. And while the pupils of Linnaeus searched the world for exotic species, as Solander did on Cook's first great voyage of discovery, nearer home naturalists began to find a wealth of unsuspected life on the sea shores.

Among the first, and greatest, of these naturalists was John Ellis, correspondent of Linnaeus and Fellow of the Royal Society of London.

In the middle years of the century he visited the shores of the Isle of Sheppey, of Sussex and of Kent in pursuit of his inquiries into those plant-animals, or zoophytes (as they were then styled), which he was to reveal as true animals. In the fourth volume of his *British Zoology*, published in 1777, Thomas Pennant provided what may fairly be described as the first straightforward account of many of our shore animals. Though far from being as full and detailed as his earlier volumes, which deal with the better-known mammals, birds, reptiles and fishes, it contains excellent descriptions and still better figures of many of our common crustaceans, which Pennant separated from the insects with which Linnaeus had grouped them, together with worms, molluscs, starfish and sea-urchins. To-day Pennant is probably best known as the recipient of letters from Gilbert White of Selborne and for the accounts he wrote of his numerous tours throughout the British Isles, but he holds a high place in the history of British zoology. Many early shore collectors must have relied on his book for the identification of their specimens.

As the eighteenth passed into the nineteenth century, and despite a Europe in arms, the study of shore life proceeded steadily. The wealthy and cultivated but somewhat pedantic eighteenth-century amateur gave place to the retired naval or India Company surgeon, to the lawyer and clergyman or retired merchant. As the century went on these in turn were gradually replaced by the professional naturalist filled with an enthusiasm that survived a meagre livelihood wrung from writing or from teaching in the new Universities and Colleges. The growth of marine biology owes little to the senior English, though much to the Scottish, Universities. It is to these men, united in little else beyond a common interest in nature, that our present detailed knowledge of shore animals and plants is largely due. Between them they had produced by the middle years of the century a series of excellent monographs, many of which we shall have reason to mention in due course, on all the more important groups of marine animals and plants.

Meanwhile great changes were taking place in the social fabric of this country. The industrial revolution had carried wealth to a numerous middle class. With the advent of railways this wealth was gladly spent seeking relief from the smoke and clatter of manufacturing cities in the clean atmosphere and peace of the seaside. The British people had begun to seek the solace of the sea shore. And almost simultaneously there appeared a naturalist who was to reveal to this essentially serious-minded generation the beauties to be gleaned—and the moral lessons to be learnt—from the study of shore life. Aided by the writings and the equally admirable drawings of Philip Henry Gosse, the Victorians came to discover the delights of shore collecting so that, quoting the words of G. M. Trevelyan in his *English Social History*, " In remote creeks and fishing hamlets, where families from town came to lodge, children and their parents bathed and dug and searched the tidal treasures of the rocks."

It is impossible adequately to write on the shore life of Britain without mentioning Gosse. To-day he survives chiefly as the father in the *Father and Son* of Edmund Gosse, although all who recollect childhood pleasure in the adventures of Tom with the lobster, with the last of the Garefowl and with Mother Carey's Chickens in *The Water Babies*, owe something to Gosse, who infected Charles Kingsley with his love of the shore and its animals. As a zoologist Gosse holds enduring reputation as the author of *Actinologia Britannica*, until recently the standard work on British sea-anemones and corals. But to the mid-Victorians his was a household name. His career in its unpredictable sequences of scene and event, but sustained throughout by an intense love of nature was typical of that of so many of his fellow naturalists. It is recounted in the *Life* which his son wrote many years before the better-known *Father and Son*.

He was born in 1810, the second son of an itinerant miniature painter, from whom he was to inherit artistic gifts, who had married a farmer's daughter many years his junior. Two years later the family settled at Poole in Dorset from whence Philip sailed when seventeen years old to enter a mercantile firm in Newfoundland. Eleven years later, after subsequent experience of farming in Canada and school teaching in Alabama, he returned to England richer only in knowledge and love of nature but with the almost completed manuscript of *The Canadian Naturalist*. Successfully published in 1840 it contained, in the words of his son, " The germs of all that made Gosse for a generation one of the most popular and useful writers of his time . . . the picturesque enthusiasm, the scrupulous attention to truth in detail, the quick eye and the responsive brain, the happy gift in direct description." Other books, honest pot-boilers in the main, were to follow, and then came eighteen months of collecting, chiefly birds, in Jamaica. He returned to still more intense writing and scientific work in London, chiefly memorable for his *Birds of Jamaica* and *Naturalist's Sojourn in Jamaica*, until shattered health compelled him (" wife, self and little naturalist in petticoats ") to seek recuperation in the sunshine of the Devon coast.

Gosse found more than health in Devonshire ; he found a scope for all his talents. The observing eye of the naturalist, the descriptive pen of the writer and the delicate brush of the artist were to find common expression in a unique series of books on shore life. In his *Naturalist's Rambles on the Devonshire Coast; The Aquarium : An Unveiling of the Wonders of the Deep; Tenby : A Sea-Side Holiday;* and *A Year at the Shore* he took his fascinated countrymen and countrywomen into a new wonderland among the rocks and beaches of their own shores. He achieved fame and, for the first time in his hard life, some measure of wealth and comfort. Charles Kingsley became one of the first and most devoted of his disciples. An article first published in the *North British Review* formed the basis of his charming *Glaucus : or, The Wonders of the Shore*, to which a *Companion* containing coloured plates with descriptions by G. B. Sowerby was added later. In *Glaucus*, Edmund Gosse writes, " the lilies of my father's praise are sprinkled with full hands." Gosse conducted shore classes at Ilfracombe, of which his son recalls memories in the *Life*. " At the head of the procession, like Apollo conducting the Muses, my father strides ahead in an immense wide-awake, loose black coat and trousers, and fisherman's boots, with a collecting-basket in one hand, a staff or prod in the other. Then follow gentlemen of every age, all seeming spectacled and old to me, and many ladies in the balloon costume of 1855, with shawls falling to a point from between their shoulders to the edge of their flounced petticoats, each wearing a mushroom hat with streamers ; I myself am tenderly conducted along the beach by one or more of these enthusiastic nymphs, and ' jumped ' over the perilous little water-courses that meander to the sea, stooping every moment to collect into the lap of my pink frock the profuse and lovely shells at my feet. This is one memory, and another is of my father standing at the mouth of a sort of funnel in the rocks through which came at intervals a roaring sound, a copious jet of exploding foam, and a sudden liquid rainbow against the dark wall of rock, surrounding him in its fugitive radiance. Without question, this is a reminiscence of the Capstone Spout-Holes, to which my father would be certain to take the class, ' the ragged rock-pools that lie in the deep shadow of the precipice on this area ' being, as he says in the *Devonshire Coast*, ' tenanted with many fine kinds of algae, zoophytes, crustacea, and medusae.' "

In such a manner to the chosen few and to the multitude through the stimulus of his writing and the beauty of the paintings and sketches of shore creatures that adorned his books, Gosse drew young and old of both sexes to the contemplation and collection of life on the shore. With no less success he induced them to transport and maintain marine animals and plants in their own homes. The credit for the invention of the marine aquarium, maintained by a nice balance between plant life and animal life, the former providing oxygen and utilising the inorganic excrement of the animals, he shared with contemporaries but he was undoubtedly primarily responsible for its wide popular adoption. In his *Handbook to the Marine Aquarium* he gave " practical instructions for constructing, stocking, and maintaining a tank, and for collecting plants and animals." He gave advice about the transport of sea water in casks, preferably of fir-wood, or in stoneware jars. But to overcome " the inconvenience, delay and expense attendant upon the procuring of sea-water, from the coast or from the ocean," he worked out a formula for the preparation of artificial sea water. Appropriate quantities of four salts only, common table salt, epsom salts, chloride of magnesium and chloride of potassium, were alone needed and he found that the majority of shore animals and seaweeds lived indefinitely in this medium. The keeping of home aquaria became immensely popular and led to the foundation of large public aquaria, the first, in 1853, at the Zoological Society of London which was initially stocked by Gosse, and later in many other towns, culminating in the grandiose Brighton aquarium with its 110,000 gallons of sea water. Still commonly on the dustier shelves of second-hand bookshops may be found copies of Gosse's *Handbook* and of many later imitations such as the one now open before the writer, *The Book of the Marine Aquarium* by Shirley Hibberd, inscribed in faded ink on the flyleaf to " Alice Hardcastle from her Mother with fond love and best wishes on her Birthday 1864. Presented with an Aquarium."

A flood of popular books on shore life sought to exploit the popular interest roused by Gosse. Few are of enduring value but some were the work of true naturalists such as G. B. Sowerby's charmingly illustrated *Popular History of the Aquarium of Marine and Fresh-Water Animals and Plants* and the lengthy and in places amusingly garrulous *Aquarian Naturalist* of Professor Rymer Jones. Also attracted to the sea shore for " philosophic inquiry into the complex facts of life " was George Henry Lewes, whose *Sea-Side Studies* breathe a very different spirit from those of Gosse. Lewes was not a naturalist but, as the biographer of Goethe and admirer of Richard Owen, discussed in the light of his findings on the shore the fundamental problems of comparative anatomy, of physiology and of generation.

As the nineteenth century progressed the Victorians turned to other things and interest in shore life waned. It was, on the whole, just as well. Emphasis throughout had been on the collection of specimens, inevitably after the introduction of aquaria, rather than on their observation in nature, and the shore collector set forth suitably equipped for his task. Lewes's description will serve. " We are thus arrayed : a wide-awake hat ; an old coat, with manifold pockets in unexpected places, over which is slung a leathern case, containing hammer, chisel, oyster-knife, and paper-knife ; trousers warranted not to spoil ; *over* the trousers are drawn huge worsted stockings, over which again are drawn huge leathern boots." Kingsley even writes of the hiring of a " strong-backed quarryman, with a strong-backed crowbar." Such methods had their natural result. After describing the wealth of beauty in the virgin rock pools of the Devonshire coast when he first viewed them with his father, Edmund Gosse in *Father and Son* proceeds with sorrow, " All this is long over, and done with. The ring of living beauty drawn about our shores was a very thin and fragile one. It had existed all those centuries solely in consequence of the indifference, the blissful ignorance of man. These rock-basins, fringed by corallines, filled with still water almost as pellucid as the upper air itself, thronged with beautiful sensitive forms of life—they exist no more, they are all profaned, and emptied, and vulgarised. An army of ' collectors ' has passed over them, and ravaged every corner of them. The fairy paradise has been violated, the exquisite product of centuries of natural selection has been crushed under the rough paw of well-meaning, idle-minded curiosity. That my father, himself so reverent, so conservative, had by the popularity of his books acquired the direct responsibility for a calamity that he had never anticipated,

---

PLATE 3

CORNISH ROCK POOL with seaweeds : *Cystoseira ericoides* (greenish-blue), *C. fibrosa* (brown tufts), *Scytosiphon lomentarius* (long strips), *Corallina officinalis* (purple), *Enteromorpha* (green)

PLATE 3

D. P. Wilson

PLATE 4

D. P. Wilson

became clear enough to himself before many years had passed, and cost him great chagrin. No one will see again on the shore of England what I saw in my early childhood, the submarine vision of dark rocks, speckled and starred with an infinite variety of colour, and streamed over by silken flags of royal crimson and purple."

Interest in marine life was maintained but now the deep seas had first claim. The triumphs of the early dredgers, who brought strange forms of life to the surface so that it seemed that the very origin of life itself might there be found, took hold on the imagination of laymen and scientists alike. The achievements of Edward Forbes and his friends off the Scottish coasts and in the Aegean Sea, and of Wyville Thomson and Carpenter in the Atlantic and the Mediterranean, culminated in the great voyage of H.M.S. *Challenger* from 1872 to 1876, and the science of oceanography was born. Into this broader synthesis of knowledge the study of shore life and of the fish and fisheries became drawn. The study of the life of the sea shore became increasingly academic as marine biological laboratories were established around our coasts.

In 1867 a young German zoologist, Dr. Anton Dohrn, visited this country in the hope of studying certain marine crustaceans. He was advised by Dr. Baird of the British Museum to seek his animals at Millport on the island of Great Cumbrae in the Firth of Clyde because there he would have the help of David Robertson, an amateur marine biologist of wide experience who had settled there after a business career in Glasgow. This visit would appear to have been of prime influence in the later foundation by Dohrn of the famous Stazione Zoologica at Naples. This great pioneer marine laboratory was erected in 1872 and became a truly international centre of research which has fortunately survived the recent war. The first British marine laboratories were established in Scotland in 1884, one, mainly concerned with fisheries, at St. Andrews and the other in a flooded quarry at Granton near Edinburgh. The latter was founded by Dr. (later Sir) John Murray who had been a member of the *Challenger* Expedition and was for many years the leading British oceanographer.

PLATE 4
RELATIVELY SHELTERED SHORE on coast of South Devon showing abundant growths, specially of brown *Ascophyllum nodosum* and *Fucus serratus* and red *Rhodymenia palmata* in the background

The laboratory consisted of a converted wooden lighter, the *Ark*, but the following year Murray had this towed through the Forth and Clyde Canal to ·Millport. There, in happy conjunction with Robertson, the *Ark* remained to become the precursor of the present laboratory of the Scottish Marine Biological Association, which was opened in 1897. The wide interest in fisheries and in marine biology generally which was stimulated by the International Fisheries Exhibition held in London in 1883 led to the foundation of the Marine Biological Association of the United Kingdom and the erection of their laboratory at Plymouth. Opened in 1888, this now ranks as the major institution of its type in Europe. Other marine laboratories established at Port Erin in the Isle of Man and at Cullercoats on the coast of Northumberland, in association with the Universities of Liverpool and Durham respectively, have been the centre of further scientific study of shore life. The work of the government laboratories at Lowestoft and Aberdeen is concerned essentially with food fishes and so lies outside the scope of this book.

The bulk of recent additions to our knowledge of shore life has come from these laboratories where continuous observations can be conducted over long periods and in which all manner of experimental studies can be made. Some of this work has been done by the permanent staffs of the institutions, still more by visiting workers from the universities of this and many other countries. There are still new species remaining to be discovered on our shores, but in the main attention has now been transferred to problems of distribution, or zoning, between tide marks, to the inter-relationships of the various members of the fauna and flora and to the study of the innumerable devices of structure and of function which enable so many creatures to live and to reproduce themselves under the constantly fluctuating conditions of life on the shore. And, most significant of all, the animals and plants of the sea shore are no longer studied as something apart but in relation to the broad problems of oceanography and against the wide background of the open seas from which they came.

CHAPTER 3

THE INHABITANTS OF THE SHORE

"On descending from terrestrial objects to the inhabitants of the waters, infinitely new and interesting matter is presented for the contemplative physiologist. Myriads of beings, alike singular in structure and properties, appear in their peculiar element, all actuated by the resistless impulse of nature ; avoiding danger, seeking subsistence, rendering the weaker a prey."

J. Graham Dalyell : *Observations on Planariae.* 1814

T H E diversity of living things upon the shore involves the writer in difficulties only solved by a brief preliminary description of the chief types of animals and plants that were introduced by Gosse to a former generation. They are our prime concern and while this volume is far from being a text-book of zoology or of botany, some account of the nature of the inhabitants of the shore must precede later descriptions of their habits and distribution. Only some of the commoner species can be shown in the plates and figures, although care has been taken to include examples of all the major groups. But for the majority, reference must be made to the numerous works, products in the main of the enthusiastic labours of nineteenth-century naturalists, which cover most types of shore-dwelling animals and plants. These are mentioned either in the text or listed in the Appendix, and copies of some of their illustrations appear here as figures. Nomenclature is a difficult problem. Throughout this book common names are given wherever possible, but in all cases the scientific names, using those given in the *Plymouth Marine Fauna* (second edition, 1931) together with older, but well-established, names in brackets where necessary to avoid confusion.

The most numerous animals between tide marks are probably Protozoa. But they are almost all excessively minute, consisting usually of a single cell, and we are not here concerned with anything not visible to the naked eye with the aid perhaps of a simple lens. The only shore protozoans of any size are foraminiferans, which usually construct a limy chambered skeleton. Certain tropical species are about the size of a silver threepenny piece while the fossil *Nummulites* attained a diameter of up to seven inches. Our largest shore form, common in rock pools, is *Gromia*, which has a simple oval skeleton, brown and leathery, about two millimetres across. Protozoa are more readily seen in mass. Green and brown patches on sandy shores or green colorations in pools high on the shore are often caused by immense aggregations of flagellate protozoans, individually among the smallest of organisms. They have characteristics of both plants and animals and from something of this kind both of these kingdoms of living things may originally have evolved. The phosphorescence sometimes seen along the margin of the sea is frequently due to swarms of the flagellate protozoan *Noctiluca*, which may be left behind as a red scum on the sand along high tide mark. The property of phosphorescence is, however, of wide occurrence among marine animals.

Leaving the protozoans we pass to larger animals in which the body is composed of many cells which are organised to form a single individual. These are known collectively as metazoans and among their many cells the various functions performed by the single cell of the protozoans are distributed. They are broadly divisible into the sponges on the one hand and the much greater mass of the remainder, called the enterozoans, or animals with digestive tubes, on the other.

Sponges (Pls. 25 and 26, pp. 144, 145) are among the most remarkable of animals. Indeed it is by no means obvious that they are animals at all and hardly surprising that they were not so recognised until comparatively recently. As inhabitants of the sea they have been known from remote times ; Glaucus who built the *Argo* in which Jason and his companions sailed in search of the Golden Fleece was a sponge diver. Aristotle in his *Historia Animalium* has much to say of them and considered them as intermediate in character between plants and animals. But the early naturalists of the modern period all included them in the plant kingdom and so did Linnaeus until John Ellis, in the face of much opposition, persuaded him to group them with the animals. Controversy continued to rage on the subject until in 1825 Dr. R. E. Grant of Edinburgh observed under the microscope " this living fountain vomiting forth, from a circular cavity, an impetuous torrent " and showed once and for all that sponges possess the fundamental property of animals by feeding on particles of organic matter and not on inorganic matter as do plants. Sponges are always attached and are best described as animated sieves drawing in water through many minute openings and, after straining off the food particles,

PLATE I

D. P. Wilson

a. Attached or Scyphistoma generation of the Common Jellyfish, *Aurelia aurita*
Photographed under water (×3)

D. P. Wilson

b. Scyphistomata in process of strobilating with formation of ephyrae
Photographed under water (×4)

PLATE II

a. Group of strobilating Scyphistomata of *Aurelia aurita* showing a newly liberated ephyra. Photographed under water (×1¼)

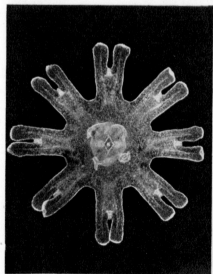

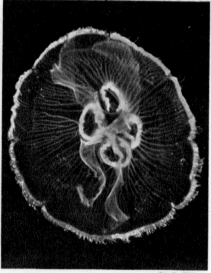

b. Ephyra of *Aurelia aurita*
Photographed under water (×12)

c. Common Jellyfish, *Aurelia aurita*
Photographed under water with under
surface uppermost (×⅓)

discharging it, as Grant so vividly described, through occasional larger apertures called oscula, which can usually be seen with the naked eye. It is important to note that they are *not* mouths ; indeed the whole arrangement of the internal cavities in sponges is totally different from that of other animals.

Sponges are common on rocky shores and various kinds will be described later. British sponges are useless commercially because they possess, like the great majority of sponges, a skeleton of delicate spicules of silica or of lime. Commercial species are confined to warmer seas, the best quality coming from the eastern Mediterranean, where they occur at some depth ; they have a flexible horny skeleton which is exposed when the living tissue rots away.

It can be said of all remaining animals, invertebrates and vertebrates, that they possess—or have possessed, because some are degenerate parasites—a mouth giving access to the gut. But not all possess a second, or anal, opening. This is absent in the radially symmetrical coelenterates which now claim attention. These creatures, which include the well-known sea-anemones and jellyfish, are all aquatic and, like the sponges, largely marine. In structure they are very simple. An outer layer of cells bounds the body, an inner layer lines the gut cavity, and between the two is a layer of jelly-like material which forms the great bulk of the substance in both anemones and jellyfish. The mouth is typically surrounded by tentacles. But simplicity of structure is no bar to success and no group of marine animals is more widely distributed.

These simple animals would seem to owe success to the possession of stinging cells and to the capacity of many of them for forming elaborate colonies. The tentacles are armed with batteries of nettle-cells (nematocysts) which protect the delicate bodies and enable them to seize as prey the animals, usually more active and sometimes larger than their captors, on which alone they feed. The stinging cells of some jellyfish are powerful enough to penetrate human skin, as many bathers have reason to know. In forming colonies the first-formed animal divides without the daughter individuals separating and, the process being continued, an elaborate branching or rounded colony is gradually produced and supported by a skeleton of lime or horny material. Reef-building corals are merely colonial sea-anemones with the power of forming massive limy skeletons.

Three types of these animals are common on our shores and in

T.S.S. C

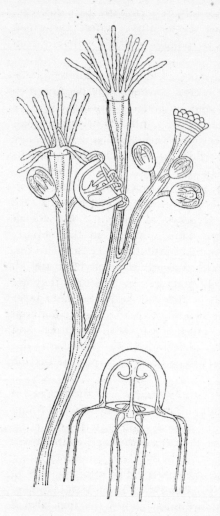

FIG. 1—Hydroid, *Bougainvillia ramosa* ; part of colony with polyps and developing medusae, also free medusa. Greatly magnified. (From Allman, *A Monograph of the Gymnoblastic or Tubularian Hydroids.*)

our tidal waters, namely hydroids, jellyfish and sea-anemones. The first are small and generally inconspicuous. They form delicate, plant-like colonies (Pls. 23 and 24a, pp. 136, 137) which bear numerous polyps, each a mouth with a surrounding ring of tentacles, like flowers on the branches. Hydroids are among the " zoophytes " that were for long regarded as part animal and part plant until John Ellis showed them to be in reality " ramified animals." All are attached and accordingly restricted, like sponges, to rocky shores where we shall find them in abundance on stones and seaweeds and sometimes on the shells of other animals.

One of the great zoological discoveries of the last century was the revelation that there is an alternation of generations in the life of many hydroids and jellyfish. During the summer many hydroids give rise, either singly or in groups within elongated receptacles, to minute jellyfish or medusae (Fig. 1). These are formed asexually, by a kind of budding, but are themselves sexual, being either male or female. Like the larger and more familiar jellyfish, they swim by pulsations of the bell, water being forced gently out at each contraction of the muscles which encircle the margin of the bell. Hence they progress with the rounded surface foremost and the mouth behind. When ripe they discharge egg or sperm into the sea and the embryos later produced sink eventually to the bottom, where they develop into the fixed or hydroid stage and the life-history is complete.

The initial discovery of this remarkable sequence of events was made in 1829 by the Norwegian pastor, Michael Sars, who later became Professor of Zoology at Christiania (Oslo). But some of the credit goes to Sir John Graham Dalyell, whose independent observations were published somewhat later in his *Rare and Remarkable Animals of Scotland, represented from living subjects : with Practical Observations on their Nature.* The author, a pioneer worker in marine biology who has never received adequate recognition, was hardly less remarkable than the animals he described. By profession a lawyer of Edinburgh, he was an eminent antiquarian and a skilled musician as well as a zoologist. We are told that he also delighted in mathematics and in delicate turning with the lathe. His artistry is revealed in many of the numerous coloured plates that illustrate the work already mentioned and his later, and partly posthumous, volumes entitled *The Powers of the Creator Displayed in the Creation ; or, Observations on Life amidst the various Forms of the Humbler Tribes of Animated Nature : with Practical Comments and Illustrations.* Much the greater part of the five large volumes which comprise these two works is devoted to marine invertebrates. He collected the majority from the Firth of Forth, which Kingsley refers to as being, under the auspices of Dalyell, the original home of marine biology in Scotland, as Torquay, where many pioneer marine biologists worked and where Gosse finally made his home, may be regarded for England.

Hydroids are divisible into two types, those in which the polyps can be withdrawn into the protection of skeletal cups and those that are naked. Not all exhibit alternation of generations ; the medusoid stage may be reduced and their sexual products produced on the hydroid. In the well-known freshwater polyp, *Hydra*, there is no trace of a medusoid, but we regard this as a consequence of life in fresh water and the apparently more complex alternation as the primitive condition.

The jellyfish are not, strictly speaking, shore animals although they are frequently found stranded between tide marks. But they are so common in shallow water as to demand attention. Moreover their small hydroid stage (Pl. I, p. 16) may be found in rock pools or on pier piles. In these coelenterates the medusoid stage is said to be dominant ; the inconspicuous white hydroid, the *Hydra Tuba* of Dalyell, now called the scyphistoma, differs from the true hydroids in being solitary, lacking a skeleton and in details of internal structure. The manner in which the jellyfish are formed from it is remarkable. As shown in the very beautiful photographs by Mr. D. P. Wilson (Pls. I and II, pp. 16, 17), the long tentacles are withdrawn while the animal elongates and becomes ringed with annular constrictions. These finally break through, thus cutting off a series of eight-lobed discs, which, as young jellyfish, swim away. The whole process is known as strobilisation. With growth the spaces between the arms of the small jellyfish or ephyrae are filled in until the gently scalloped outline of the adult bell is attained. The whole process of development in the common jellyfish, *Aurelia aurita*, is shown in the Plates. Meanwhile the base of the hydroid individual is reorganised, new tentacles grow out and the animal resumes feeding and growth, and becomes eventually able to produce a new batch of ephyrae.

There are so few jellyfish in our waters that all may be mentioned (Fig. 2, p. 21). *Aurelia* is the commonest and often extends into estuaries and docks. It is interesting as being of almost world-wide distribution. Near the centre of the flat bell lie four purple rings marking the site of the reproductive organs. The bell is seldom more than about one foot in diameter ; its margin is frilled with many small tentacles and bears eight sense organs which are influenced by the force of gravity and by the stimulus of light. The borders of the mouth are prolonged into four arms used in feeding. The threads from the sting cells of *Aurelia* seldom penetrate human skin but this is not true of those of the larger *Chrysaora isosceles*, easily distinguished by the bands of white and brown that radiate out from the centre of the disc and by the 24 long tentacles trailing from the margin. The two species of *Cyanea*, the blue *C. lamarcki* and the yellow *C. capillata*, also sting. They carry 8 tufts of long tentacles and the margin of the bell is deeply lobed. The largest of our jellyfish, but as harmless as *Aurelia*, is *Rhizostoma pulmo*, which is altogether more massive than the others and can attain a diameter of some 2 feet. There are no marginal tentacles but the oral arms are very long and sub-divided. They fuse over the mouth opening but are perforated with many minute apertures through which are passed the small animals on which they feed. The bell is pale blue or green with a border of deeper colour.

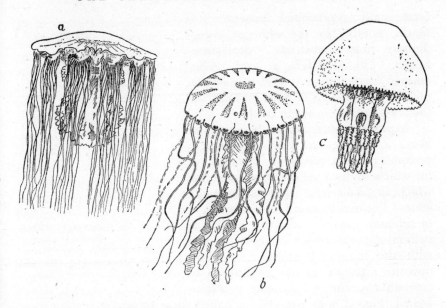

FIG. 2—Common British Jellyfish ; *a. Cyanea* ; *b. Chrysaora* ; *c. Rhizostoma*. Variously reduced. (Original drawings by Mr. F. S. Russell.)

Much the most obvious of shore coelenterates are the relatively highly organised sea-anemones (Pls. 8, 9, 10, pp. 81, 84, 85). These anthozoans, or flower-animals, never have a medusoid generation and the gut cavity is subdivided by radially disposed partitions which make for greater efficiency in digestion. Tentacles are numerous, usually in multiples of six. There are no more beautiful animals in the sea than fully expanded anemones, in which grace of form is united with beauty of colour and pattern. The interested observer should consult the coloured plates in Professor T. A. Stephenson's *British Sea Anemones* for a permanent record of the necessarily transitory impressions of the sea shore. Sea-anemones are commonest on rocky shores but do also occur in sand and mud.

True, or stony, corals are rare around British coasts, but we have two simple, or cup, corals (Pl. 11, p. 92) in which the essential resemblance to sea-anemones is clear.

A second major division of the anthozoans contains animals with 8 pinnate tentacles to each polyp. These alcyonarians are only

occasionally represented between tide
marks, notably by the white or orange
digitate mass known as " dead-man's
fingers " (Pl. XXVIIIa, p. 175). Allied
animals are common at moderate depths,
notably the sea-fans of hard bottoms and
the sea-pens which live in mud.

A passing word should be said here
about a small group which superficially
resemble jellyfish, namely the ctenophores,
for which we have no common name but
which the Americans refer to as comb-
jellies. The usually ovoid body bears eight
meridional rows of vibratile combs or
swimming-plates which drive the animal
with the mouth foremost, i.e. in the
opposite direction to the jellyfish. Two
ctenophores, the sea-gooseberry, *Pleuro-
brachia* (Fig. 3), and a larger one called *Beroe* are sometimes common

FIG. 3—Sea-gooseberry, *Pleuro-
brachia pileus*, showing ctenes or
swimming plates characteristic
of the Ctenophora, also two
long tentacles. Natural size.

close to the shore and may be found in rock pools at low tide. The
iridescent ripple of colours along the lines of swimming-plates renders
them amongst the most beautiful of marine creatures and they are on
occasion a source of phosphorescence.

Leaving these radially symmetrical animals we turn next to the
worms, " the enliveners of wet places " in Pennant's delightful phrase.
In his day all long, crawling invertebrate animals were included within
a comprehensive group of Vermes. All of them possess a head bearing
sense organs with which they explore the ground over which they
crawl ; they have thus an anterior and a posterior end and also upper
and lower surfaces and are bilaterally symmetrical. But many fundamental differences in structure have been revealed beneath the
superficial resemblance of the worm-like form and crawling habit.

The simplest are the flatworms ; many are parasitic but one group
is free-living and representatives live on the shore (Pl. XXIVa,
p. 159). But the largest is only half an inch long and so thin that it
seems to flow like a film over the surface of the rocks. It is very
difficult to detect flatworms in their natural surroundings but they will
crawl out of collections of weeds or from scrapings made from rocks
after these have been placed in a bowl of sea water. The flattened,

leaf-like form and the gliding motion are unmistakable but some magnification will be needed to reveal that the mouth lies about mid-way along the under-side of the body and that there is no anal opening.

The thread-worms (nematodes) are ubiquitous ; they live in the sea, in fresh water, in soil and as parasites in both animals and plants. Few animals are so common and at the same time so inconspicuous. They are round, pointed at both ends with the mouth at one end and the anus near the other, and are enclosed in a firm cuticle. They are singularly devoid of external organs and easy enough to identify as nematodes but usually most difficult to separate into species. Although there are many common shore nematodes they need not detain us here because all are small and inconspicuous. This is not true of the nemertine worms which include the well-known bootlace-worm, *Lineus* (Pl. XIXb, p. 142), the intertwining body of which may be yards long. These worms are strange creatures. They show greater complexity than the flatworms in possessing an anus, and than the nematodes in the development of a true blood system. But they have acquired length without a satisfactory means of controlling it. They are carnivorous and have a long proboscis which is shot out when they feed. In some the end of this is armed with spines which can be replaced if damaged from a set of "spares" conveniently carried for this purpose.

The great majority of the shore worms are annelids. These have achieved length but also solved the problem of its adequate control by having the body subdivided into segments each with its own set of internal and external organs. All are under the general control of the head region, which is better developed than in other worms and usually carries a series of tentacles having powers of touch, smell or taste, and often simple eyes. Movement in these animals is brought about by nervous impulses transmitted from segment to segment with consequent contraction of appropriate muscles.

These segmented worms are a highly successful group with a wide diversity of form and habit. The common ragworms (*Nereis*) are typical examples of the bristle-worms (polychaetes—many-bristled worms) in which all the body segments carry lateral outgrowths called parapodia with projecting bristles (Fig. 43, p. 150). With their aid the animals crawl actively, while enlarged parapodia in some are used for swimming. A second group of bristle-worms live in tubes (Pl. 37a, p. 244) ; they have no organs of locomotion but the head bears a crown of tentacles for feeding and respiration (Fig. 4, p. 24). They rely for protection on sudden with-drawal into the shelter of the tube, which is formed from a variety of materials in different species.

The terrestrial earthworms have few counterparts on the shore. They belong to another group of annelids without para-podia but with a complicated hermaphrodite reproductive system. These oligochaetes (few-bristled worms) are abun-dant in fresh water, but shore forms are small and need to be carefully sought. The parasitic leeches belong to yet another, but allied, group of annelids but although there are marine species which live on fish these are not normally found on the shore.

From consideration of these annelids we pass naturally to that of the largest group in the animal kingdom, the arthro-pods. These are also segmented but the simple lateral projection

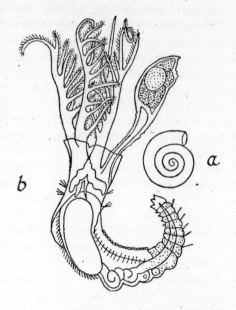

FIG 4—Serpulid tube-worm (polychaete), *Spirorbis borealis* ; *a.* calcareous coiled shell, enlarged ; *b.* animal, ×50. The tentacles are used for feeding and respiration, apart from the club-shaped one which forms an operculum, closing the opening of the shell, while the end serves as a brood-pouch for the developing eggs. (From *The Cambridge Natural History*, after Claparède.)

of the annelid is here an elaborate, jointed limb, while the thin cuticle which protects the body of the worm and stretches as this grows is replaced by a stout integument composed of a tough substance called chitin which is often further strengthened with horny or limy material. Such a covering cannot stretch with growth and hence the arthropods must moult from time to time. The old shell is cast and the new one, still soft, extends before hardening and so the animals grow in a series of jumps and not continuously like the worms.

Insects are the best known of the arthropods, indeed in wealth of species they exceed the rest of the animal kingdom put together, but they are air-breathing and, although many do live in fresh water, purely marine species are rare apart from a number of small species which live between tide marks and scavenge in the rotting seaweed and other debris along the strand line. Centipedes and millipedes are confined to the land and so are most of the arachnids, which include spiders, mites, ticks and scorpions, although the remarkable and archaic king-crab (*Limulus*) is marine and extremely common along the Atlantic shores of America where it comes inshore to lay its eggs on sandy beaches. There is an interesting little group of so-called sea-spiders (pycnogonids) with representatives on our shores (Fig. 5). They are inconspicuous little animals with four pairs of legs like true spiders and are usually found attached to sea-anemones or to hydroids, on which they feed.

The members of the remaining major group of arthropods are typically marine and are abundantly represented on our shores. These are the crustaceans, which Thomas Pennant separated from the insects with which they had previously been grouped. They range in size from the extremely minute to the impressive bulk of large crabs and lobsters and display a fascinating diversity of form and habit.

The smallest crustaceans are the copepods which are amongst the commonest of all animals. They occur in countless millions in the surface waters of the sea and are the most important members of the animal plankton which is discussed in the next chapter. A sample of water from the shore, unless taken in mid-winter, is certain to contain copepods, but they can only be adequately studied under the microscope. Some live in the water held between the grains of sand on beaches, and these are truly shore animals.

Probably the most numerous animals of any size on a rocky shore are the acorn-barnacles, which often form a dense zone along its upper limits (Pl. XIIIb, p. 114). They are cemented by a broad base to the rock surface, unlike the well-known ship's barnacles which have long stalks and attach themselves to all manner of floating objects, to buoys and floating logs as well as to ships. The animal is much the same in both and is enclosed within a series of limy plates. There is no obvious resemblance between creatures like this and typical crustaceans

FIG. 5—Sea-spider, *Pycnogonum littorale*. Twice natural size. (From Newbigin, *Life by the Seashore*.)

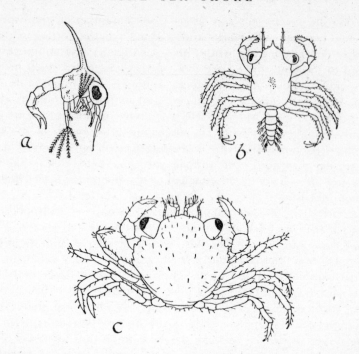

FIG. 6—Stages in the development of the shore-crab, *Carcinus maenas*. *a*. Zoea larva, shortly after hatching; *b*. later, megalopa larva with abdomen still extended; *c*. young crab showing adult form. *a*. and *b*. twenty-two times natural size; *c*. eleven times. (From Calman, *Life of Crustacea*, partly after Williamson.)

such as prawns or crabs. Barnacles were for long ages a great mystery to men. One of the most interesting of medieval myths grew up around them, namely that the stalked barnacles (*Lepas anatifera*) grew on trees and engendered what are still known as barnacle-geese. The myth took many forms which the interested reader will find recounted in E. Heron-Allen's *Barnacles in Nature and in Myth*. The mystery of the true nature of barnacles survived these myths. Even the great Cuvier was baffled; he noted the limy plates which open and close like the shell-valves of a bivalve mollusc and so classified barnacles with molluscs.

The truth came from the unexpected person of John Vaughan Thompson, an army surgeon at that time stationed at Cork where he was deputy inspector-general of hospitals. He devoted his leisure to

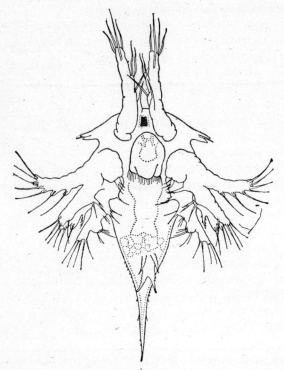

FIG. 7—Nauplius larva of the acorn-barnacle, *Balanus balanoides*. Greatly magnified. (From Calman, *Life of Crustacea*, after Hoek.)

the study of marine life and in 1830 published obscurely at Cork his *Zoological Researches and Illustrations*, now recognised as a classic work. He traced the development of the common shore-crab from hatching to maturity. He found the early larval stages to be widely different from the adult and to live in the surface waters like the copepods. With successive moults came striking changes in form (Fig. 6, p. 26), the elongated " tail " finally tucking under the rounded body and the animal, now obviously a crab, taking to life on the bottom. Thompson then proceeded to a similar study of barnacles. Again he found that the early stages, now known as nauplius larvae (Fig. 7), live free and active in the sea and, most significant of all, that they are typical crustaceans. The final change in form, or metamorphosis, was even more striking than in the crabs because the young barnacles cemented themselves to a hard surface, developed a protective casing of plates and then employed the appendages (Pl. XIIIa, p. 114), used by other crustaceans for locomotion, as a kind of cast net for collecting finely divided food-particles from the water.

The barnacles have the high distinction of being the one group on which Charles Darwin wrote a detailed monograph. He laboured on them for eight years, and though he found his self-imposed task often wearisome he regarded the knowledge so obtained of comparative anatomy and of classification as a fundamental part of his training as a naturalist. Barnacles, or cirripedes to give them their scientific name, occur also on the bodies of large marine animals such as whales and turtles and, possibly from some such initial habit, others have passed to parasitism within the bodies of larger crustaceans.

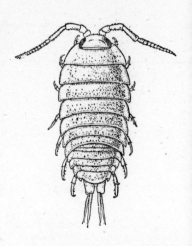

FIG. 8—Sea-slater, *Ligia oceanica.* Twice natural size. (From Calman, *Life of Crustacea,* after Sars.)

There are a large number of small shore crustaceans from about a quarter of an inch to an inch long. The three commonest groups are the isopods, amphipods and opossum-shrimps or mysids. The first are flattened from above to below and include the well-known terrestrial wood-lice or slaters. The commonest shore species is the relatively

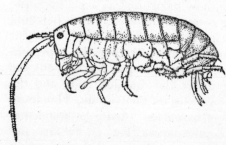

FIG. 9—Common sand-hopper, *Talitrus saltator*, male. Three times natural size. (From Calman, *Life of Crustacea,* after Sars.)

large sea-slater, *Ligia oceanica* (Fig. 8), which lives in cracks in rocks about high tide level. The amphipods are also flattened, but from side to side. They are extremely numerous and include the hoppers (Fig. 9) which appear sometimes almost like a cloud when stranded seaweed is disturbed or rocks turned. The opossum-shrimps are more nearly allied to the

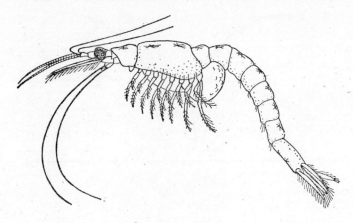

FIG. 10—Opossum-shrimp, *Praunus flexuosus*, female showing brood-pouch projecting below and behind the thorax. Magnified. (From Ritchie, *Thomson's Outlines of Zoology*.)

true shrimps and prawns and have the same body form (Fig. 10). They have appendages adapted for swimming and not for crawling, jumping or burrowing like the isopods and amphipods, and may be found in pools during the summer months. All these animals have in common a brood pouch formed by overlapping plates on the under-side of the body and there the young develop to leave the female as miniatures of the adult. There is no larval stage and no striking metamorphosis in development as there is in the more lowly barnacles and the more complex shrimps, prawns and crabs.

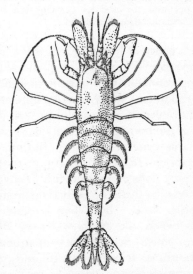

The larger shore crustaceans are known as decapods because they possess five pairs of walking legs (these including the large claws of lobsters and crabs). This group comprises many animals with a great range in form and a corresponding

FIG. 11—Common shrimp, *Crangon vulgaris*. Natural size. (From Calman, *Life of Crustacea*.)

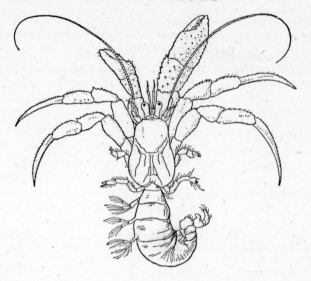

FIG. 12—Hermit-crab, *Eupagurus bernhardus*, removed from shell to show the soft asymmetrical abdomen and the terminal sickle-shaped appendages which grip the central pillar of the shell. The other abdominal appendages in this, a female, specimen are used for carrying the developing eggs. About natural size. (From Calman, *Life of Crustacea*.)

diversity of habit. First, there are the swimming or natant forms, the prawns (Pl. 12a, p. 93) and shrimps (Fig. 11), not unlike the small opossum-shrimps in general appearance and, like them, having a long abdomen,[1] which carries appendages used for swimming. Normally, however, the intertidal prawns spend most of their time moving about on the bottom of rock pools, while the shrimps burrow in sand. Of the same general body form but much larger are the " crawling " (reptant) lobster (*Homarus*) (Pl. 13, p. 100) which is widely distributed, and the spiny lobster or crawfish (*Palinurus*) which is confined to the south-west. In these the body is much stouter and the shell greatly strengthened with lime. Both live around rocky coasts and are not common between tide marks while the allied Norway lobster or marine crayfish (*Nephrops*), which is not infrequently offered for sale, comes from muddy bottoms at some depth and never appears on the shore. These animals do not swim like prawns but they can dart backwards through the water by sudden flexions of the abdomen and the tail.

[1] Often called the " tail," though strictly not so.

The next subdivision of the decapods is difficult to describe briefly because it comprises a number of apparently dissimilar animals with the fundamental similarity much obscured. In all the abdomen is to some extent modified. The best known and most obvious of these anomurans are the hermit-crabs (Fig. 12) with a soft, asymmetrical abdominal region which is inserted for protection within the empty shell of a marine snail. Others include the charming and very suitably named squat lobsters (Pl. 14, p. 101)—stout-bodied animals, sometimes brightly coloured, with the broad abdomen carried tucked under the body, although it can be extended and used for sudden backward movements as in the true lobster. Then there are a number of burrowing forms rather like prawns but with a broad and somewhat clumsy abdomen and, finally in this mixed array, others with the typical crab-like or brachyuran form in which the reduced abdomen is kept permanently tucked under the rounded thorax. Such anomurans are represented in our shore fauna by the porcelain-crabs (*Porcellana*) (Fig. 13), which are widely distributed. The impressively handsome northern stone-crab (*Lithodes*), which resembles a large spider-crab in all but the very asymmetrical tail is, our finest example, but it must be viewed in aquaria or museums.

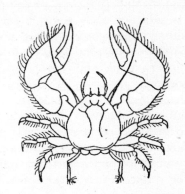

FIG. 13—Hairy porcelain-crab, *Porcellana platycheles*, natural size. (After Bell, *British Stalk-eyed Crustacea*.)

The true crabs have a symmetrical tail and are amongst the most successful and enterprising of animals. They occur in all parts of the sea, even living in the floating *Sargassum* weed of the Gulf Stream ; they extend into rivers and, in the tropics, in company with hermit-crabs, have successfully invaded the land. The crab population of our shores includes many types ranging in size from minute pea-crabs to the large edible crab (Pl. 28b, p. 153), and in form from spider-crabs (Pl. XXIIa, p. 155) with attenuated legs to swimming-crabs (Pl. 29, p. 162) in which the last pair of these end in flattened paddles.

The molluscs form as large and characteristic a group as the crustaceans. The beauty and variety of form of their shells early attracted the attention of naturalists and the study of conchology was a favourite nineteenth-century occupation as the many, often beautifully illustrated, books published at that time bear witness. It is impossible to list even the more important conchologists who contributed to the building-up of our present knowledge of British species but mention should be made of the two standard works, *A History of British Mollusca* by Edward Forbes and Sylvanus Hanley and *British Conchology* by John Gwyn Jeffreys, which, although written in the middle of the last century, still retain their value.

Molluscs are built up on a unique plan in which segmentation plays no part. Imagine a simple worm in which the internal organs became accumulated in a hump on the back and this covered with a " mantle " of tissue which formed a protective shell. You have then some idea of the manner in which molluscs may have evolved. Next visualise a " mantle cavity " at the hind end of the hump between the mantle and the underlying tissues and there is the respiratory chamber housing the gills and also, strangely enough, the anus and the openings of the kidneys and reproductive system. From some such primitive mollusc with head in front, long muscular " foot " on which it crawled like a modern snail, and above this the visceral hump with its protective shell, the many widely diverse modern molluscs are deemed to have descended. These range from conical limpets (Pl. XIVa, p. 115), adhering tightly to rocks, to more active snails with coiled shells and to bivalved cockles (Fig. 68, p. 229) burrowing in sand, and finally reach the summit of evolution without segmentation in the highly complex octopus, cuttlefish and squid. The squids are inhabitants of the surface waters, streamlined and, over short distances, the speediest animals in the sea—far removed indeed from their sluggish, bottom-living ancestors.

The shore may yield examples of four out of the five groups into which the molluscs are divided. The commonest are the marine snails with univalve shell and typically crawlers on a broad foot, hence their scientific name of gastropods or stomach-footed. The shell is typically coiled spirally but this may be changed during development, as in the limpets. The coiling is usually right-handed forming a dextral spiral, i.e. the opening is on the right when facing the observer. Left-handed, or sinistral, shells are rare ; they occasionally turn up as " sports " in normally dextral species, although a few snails, none of them British, are normally sinistral. The best known example is the sacred chank

PLATE III

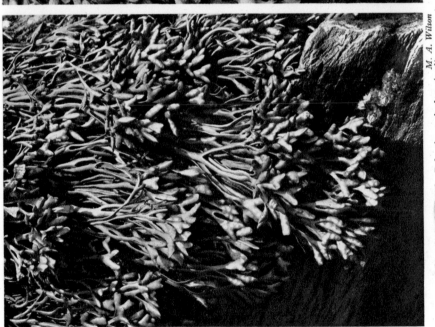

M. A. Wilson

b. Flat Wrack, *Fucus spiralis* (×⅓)

M. A. Wilson

a. Channelled Wrack, *Pelvetia caniculata* (×½)

PLATE IV

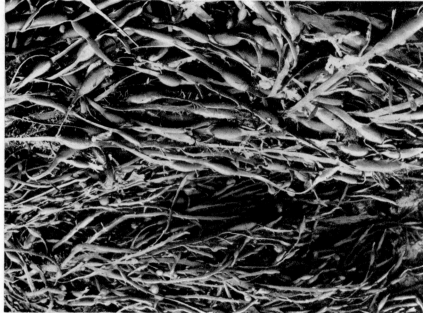

b. Knotted Wrack, *Ascophyllum nodosum* (×¼)   M. A. Wilson

D.P. Wilson

a. Bladder Wrack, *Fucus vesiculosis* (×⅓)

(*Turbinella rapa*) of southern India, usually shown held in the right hand of the Hindoo god, Vishnu, and formerly represented on the stamps of the State of Travancore. Most shelled gastropods except limpets can withdraw the head and foot within the protection of the shell by contracting a muscle attached to the central pillar or columella of the shell. In marine, but not in most terrestrial or freshwater, snails the opening of the shell is then closed by a horny or limy plug, the operculum, which is carried on the hind-end of the foot. The animal emerges as a result of hydraulic pressure, the muscle relaxes and blood is forced into the head and foot. Limpets achieve similar protection by pulling the shell firmly down against the surface of the rock. Except in the common limpet where they encircle the foot, the gills will be found in front of the hump, i.e. immediately behind the head when this is extruded. This is the result of a remarkable process of twisting round, or torsion, of the hump during development.

During evolution there has been a reduction and in some cases a loss of the shell in some marine gastropods known as the sea-slugs (Pls. 15 and 16, pp. 108, 109). We shall encounter various examples of these graceful and beautifully coloured animals in the course of our examination of the shore.

The chitons, or coat-of-mail shells, although inconspicuous, are very common shore animals (Fig. 14). The elongated body is flattened and bears eight articulated shell plates which enable the animal to bend in conformity with the irregularities of the rock surface on which it invariably crawls. In other respects they are the most primitive of existing molluscs. The bivalves, such as mussels, oysters and cockles, are well known. The laterally compressed body is entirely enclosed by the two hinged shell valves. The two valves are united by an elastic ligament which causes them to separate, or gape, when they are not being actively pulled together by contraction of the adductor muscles which run from one valve to the other. Within the now greatly enlarged gill cavity are enormous gills which have become organs of

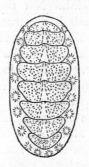

FIG. 14—Coat-of-mail-shell or Chiton, *Acanthochitona crinitus*, showing eight shell-plates and groups of spicules in surrounding girdle. Three times natural size. (From Boulenger, *Animal Life by the Sea-Shore*.)

feeding as well as of respiration. When the valves are separated a continuous stream of water is drawn in at the hind end. This passes through the fine lattice work of the gills and leaves behind, as on a sieve, the fine particles and microscopic organisms which it has carried in suspension (Fig. 38, p. 133). The water then passes out, also at the hind end but above the area where it has entered, while the food particles are conveyed to the mouth. The foot is also typically compressed laterally and is most often used for pulling its owner slowly through the sand or mud in which the majority of bivalves live.

The final group of the molluscs includes the octopus (Pl. 17, p. 122), cuttlefish (Pl. XXXIa, p. 240) and squids, all with great eyes, carrying numerous and powerful tentacles armed with suckers, and possessed of great strength and agility. They are not common on our shores although occasionally a small octopus or a stranded cuttlefish may be found at low tide, but the major evidences of the latter are the stout cuttlefish " bones," or bunches of their eggs like small black grapes attached to seaweed, washed up among the flotsam on the strand line. The bone is the strengthening skeleton of the broad body which is absent in the bag-like octopus and more slenderly developed in the torpedo-shaped squids. But fishermen lifting crab or lobster pots may bring in an octopus, and a seine net pulled over a sandy bottom may capture cuttlefish. Not the least interesting feature about these fascinating animals is the transformation of the gill cavity into an organ of jet propulsion. Powerful muscles in the bounding wall contract to expel a powerful jet of water through a narrow funnel which represents the greatly modified foot. The funnel can be turned in any direction and the animal moves in the opposite direction to that in which the water is expelled.

In originality of structure no group can compare with the starfish and allied spiny-skinned animals or echinoderms. Here we again encounter the radial symmetry of the coelenterates, with which the echinoderms were once united under the common name of radiates. But in this symmetry lies the only resemblance ; the echinoderms are far more complex and have indeed some remote affinity to the stock from which the vertebrates and finally man himself have evolved.

In fossil-bearing rocks evidence of many types of echinoderms is found but only five groups exist to-day : four of these contain a wealth of species, including some which are among the commonest of shore animals. The common red starfish, *Asterias rubens* (Pl. 31,

p. 166), forms the best starting point for a general description. This is normally five-rayed with the mouth on the under and the anus on the upper side, both centrally placed. The body is strengthened with limy plates to which are attached the characteristic spines and, scattered between these, small stalked forceps or pedicellariae which keep the surface clean. The most original feature of the internal structure found in all echinoderms is a series of tubes forming the water vascular system. This communicates with the sea water, to which medium all echinoderms are confined, by way of a perforated plate known as the madreporite (mother of pores) situated near the anus on the upper surface. Fluid consisting largely of sea water is carried to all parts of the body, including rows of small tubes which project on either side of grooves which run along the under surface of the arms. These tube feet can be distended by hydraulic pressure and bear terminal suckers. By their aid the animal makes laborious progress over any surface which these suckers can grip.

The brittle-stars (Pl. XXIVb, p. 159) are superficially very similar. They have always five arms but these are thinner than those of starfish and do not merge into the central disc, which is always sharply demarcated. The tube feet are reduced and the animals are carried along with some speed by the active, writhing movements of the spinous arms. Although extremely abundant in shallow water, brittle-stars are neither so common nor so obvious on the shore as starfish but can usually be found near low tide level or in pools.

The almost spherical shell or test of the common sea-urchin, *Echinus esculentus* (Pl. 33, p. 168), is a well-known object. Here the double rows of tube feet run like ten meridians of longitude round the body, which is supported by tightly fitting skeletal plates so that it has become rigid. The spines are long, in some tropical species longer than the diameter of the test, and move freely on rounded sockets, while the tube feet can extend for considerable distances. The mouth is armed with five teeth which are carried and operated by a complicated apparatus of struts and muscles which projects into the interior of the test and is known as Aristotle's lantern. Aristotle is fittingly so remembered because his *Historia Animalium*, the first great textbook of zoology, contains a very good account of the structure and habits of *Echinus* which he named with the Greek word meaning hedgehog.

The skeleton of a sea-urchin is a most beautifully elaborate structure. We may most fittingly describe it in the words of Edward Forbes, whose *History of British Starfishes*, published in 1841, is one of the most charmingly written and certainly the most delightfully illustrated of all books on British marine animals. Of the ambulacral regions where the tube feet project, he writes, " In a moderate-sized Urchin I reckoned sixty-two rows of pores in each of the ten avenues. Now, as there are three pairs of pores in each row, their number multiplied by six, and again by ten, would give the great number of 3,720 pores ; but as each sucker occupies a pair of pores, the number of suckers would be half that number, or 1,860. The structure in the Egg-Urchin is not less complicated in other parts. There are 300 plates of one kind, and nearly as many of another, all dove-tailing together with the greatest nicety and regularity, bearing on their surfaces above 4,000 spines, each spine perfect in itself, and of a complicated structure, and having a free movement on its socket."

Urchins of another type, with fragile tests and spines used in burrowing, will be encountered when we pass from the rocky shores that harbour *Echinus* to the soft medium of sandy beaches.

The sea-cucumbers or holothurians (Fig. 15) have elongated bodies and move with the mouth foremost. Some have tube feet but these are largely confined to the three rows on the (topographically) under-surface on which the animal crawls. Others burrow in sand, making progress through this by gripping the surface with small anchor-shaped spicules carried on the tough body. In none is there a firm skeleton, only scattered plates that take the form of wheels, anchors or anchor plates. The mouth is surrounded by a ring of feeding tentacles. In shallow tropical seas such animals are large and

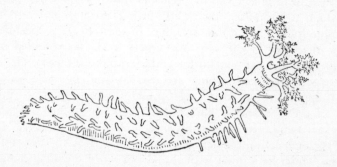

FIG. 15—Sea-cucumber, *Cucumaria saxicola*, natural size. (Modified from the *Plymouth Aquarium Guide*.)

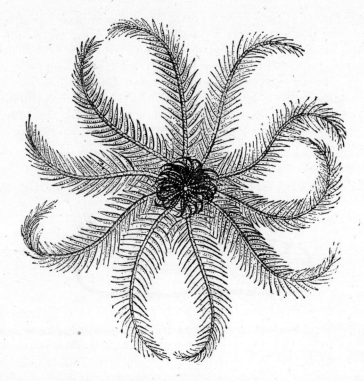

FIG. 16—Rosy feather-star, *Antedon bifida*, viewed from under surface and showing the cluster of short stalks on which it rests as well as the ten feather-like arms. Natural size. (From Forbes, *History of British Starfishes*.)

numerous and some constitute the trepang or bêche-de-mer which, after drying, are esteemed highly as food by the Chinese. There are large species at moderate depths around our coasts but the intertidal species are small and inconspicuous.

There remains our sole representative of the most primitive group of existing echinoderms, the crinoids or feather-stars. *Antedon bifida* (Fig. 16), the rosy feather-star, grows attached by a stalk but in due course breaks away. It has ten arms, each feathered with many side branches, and rests on some 25 short stalks that grow down from the central disc. It moves with languid grace by undulating movements of these arms.

There is a numerous group of animals which are often difficult to distinguish from hydroid coelenterates although they are much more highly organised. They grow in plant-like colonies made up of many minute individuals and were formerly included among the zoophytes. Vaughan Thompson, whose investigations covered a wide field, revealed their fundamental difference from hydroids and named them polyzoans, meaning "many animals." About the same time German zoologists came independently to the same conclusion but chose the name bryozoa, or "moss animal." Both names persist, together with the common name of sea-mat which is applied to those forming encrusting growths.

FIG. 17—Polyzoan, *Membranipora membranacea*, showing two compartments or zooecia, with the animal withdrawn into one and protruded from the other. Greatly magnified. (Modified after Harmer.)

The colonies are supported by well-compacted skeletons consisting of many box-like cavities each occupied by an animal (Fig. 17), which, unlike a polyp of a hydroid colony, has little connection with its fellows. In some the skeleton is strengthened with lime but it usually has a horny consistency. Each animal has a fully formed gut with mouth and anus near to one another but with the former surrounded by a U-shaped crown of tentacles for collecting finely divided food from the water. When these tentacles are protruded the surface of the colony becomes covered as with a fine white moss but any unusual disturbance, or exposure by the falling tide, causes each animal to withdraw within the protection of its particular compartment or zooecium. Some of these polyzoans form encrusting growths on the surface of seaweeds (Pl. XVIIIb, p. 139) or rocks, others grow as branching colonies. Those that live on the shore are all small and inconspicuous but some from deep water, which may be cast up on the shore after storms, are large enough to be mistaken for seaweeds.

The vertebrates, which constitute the highest group in the animal kingdom, belong to the still larger assemblage known as the chordates. These consist of animals which at some stage in the life history (in some throughout life) possess the primitive axial rod called the notochord. In the vertebrates, including man, this makes brief early appearance, to be replaced later by the segmented backbone of vertebrae. The simpler chordates never possess vertebrae and they include, in the tunicates or sea-squirts, one group which is extremely common on the shore. No animals could look more unlike vertebrates. They are bag-like creatures, always attached, with a tough surrounding tunic with two openings, through which water is squirted when the animal contracts or is squeezed (Fig. 18). In normal functioning, water passes in through one opening into a large, much-perforated sac, very like fine muslin, where it is strained, essentially as in a bivalve mollusc, and then leaves by way of the second opening. The food particles retained in the sac, really the enlarged anterior region of the gut, pass into the stomach for digestion. Such an animal, with the simplest of nervous and blood systems, is far removed from the complexity of a vertebrate. There is, moreover, no trace of notochord or any form of skeleton. Exactly as with the barnacles, the true nature of the sea-squirts is revealed only by a study of development. The egg develops into a minute " tadpole " possessing a notochord, elongated nerve cord and other typical structures of a true chordate. There can be no doubt as to the group to which this larva belongs. But then comes settlement

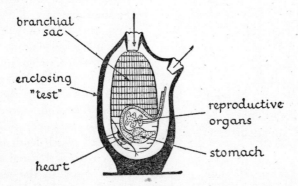

FIG. 18—Simple sea-squirt, diagram showing details of internal structure with arrows indicating inflow of water through the oral, and outflow through the atrial aperture.

on a hard surface and with it a metamorphosis not into something more elaborate, like the emergence of a butterfly from the cocoon into which the caterpillar passed, but, by a retrogressive change which involves the loss of the notochord and nerve cord, into an adult of much simpler structure. Nevertheless, like many parasites which have undergone a still greater simplification, the sea-squirts are a highly successful group, as revealed by the ultimate test of widespread abund-ance. There are few commoner animals on rocky shores, on stones among mud and down to moderate depths in the sea, from which enormous numbers may on occasion be dredged.

Three types of sea-squirts live on the shore (allied creatures live in the surface waters of the sea but they do not concern us). First there are the simple ascidians which may be a few inches long and are usually solitary, although the commonest, the gooseberry sea-squirt, *Dendrodoa* (*Styelopsis*), grows in clusters. Then there are colonial types (Fig. 19), with the individuals united at the base ; and finally the compound sea-squirts (Fig. 20, p. 41) which grow as gelatinous incrustations on the surface of rocks or on weeds (Pl. 26, p. 145). In these the individuals are very small and arranged in groups, in the form of a star or long oval, within the common test. Each individual has its own mouth opening but the second or atrial opening is common to the group. Such colonies are among the commonest objects on the shore.

As a connecting link between the sea-squirts and the fishes, although only very rarely to be encountered in the sand at the lowest ebbs, is the small lancelet, *Branchiostoma* or, using the older and much better-known scientific name, *Amphioxus*. It is some two inches long and looks at first sight like a small fish. But closer study reveals the absence of paired fins and, more primitive still, the absence of jaws. The rounded mouth leads into a wide cavity with perforated walls which strains water in the same way as the branchial sac in the sea-squirts. A notochord runs the length of the body and above this

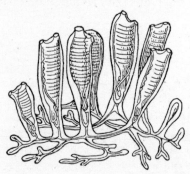

FIG. 19—Colonial sea-squirt, *Clavelina lepadiformis*, showing group of individuals arising from a common creeping stolon. Natural size. (From *The Cambridge Natural History*.)

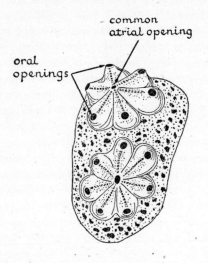

common
atrial opening

oral
openings

FIG. 20—Compound sea-squirt, *Botryllus schlosseri*, showing two groups or "systems." of individuals within common gelatinous matrix. Greatly magnified. (From *The Cambridge Natural History*, after H. Milne-Edwards.)

a nerve cord ; the creature is fundamentally like the tadpole of the sea-squirts but burrows in sand instead of swimming in the sea.

It is not necessary to say much here about the fishes. In them the notochord is replaced during development by cartilage or bone. Blocks of these arise in successive segments of the body and so the vertebrae are formed. True fish always possess jaws, a cranium to protect the brain and two sets of paired fins. The marine lampreys, which are occasionally caught in tidal waters, such as the upper reaches of the Bristol Channel, though vertebrates, are not fish because they lack jaws and also paired fins. There are two main groups of fish, one with a skeleton of cartilage the other with one of bone. The former include the round-bodied dogfish and sharks and the flattened skates and rays. They seldom appear between tide marks in these latitudes although common on the surface of tropical coral reefs. Our shore species are all bony fish. Many are typical members of the shore fauna and are variously specialised for life between tide marks.

Of the higher vertebrates, amphibians, such as frogs and newts, do not occur on the shore. They have soft, permeable skins (through which they breathe) and contact with sea water is fatal to them. Reptiles are similarly absent from our shores although in the tropics there are marine lizards and sea-snakes. Birds wander freely on all shores. Gulls of all kinds, the delightful oystercatcher, waders such as turnstones and dunlin, the little rock-pipit and the tall heron, with ducks and geese, are all to be seen on the sea shore. They demand fuller and more expert treatment and will be described in other volumes. Mammals often come on to the shore. Rats are common about high tide level where they feed on the refuse on the strand line; shrews are often found, while otters may make their homes on the shore during winter months when they feed on crabs and smaller crustaceans as well as on fish.

There are six species of seals known from the North Atlantic and of these all but one have been seen around our coasts. But the only common ones are the harbour or common seal, *Phoca vitulina*, and the grey seal, *Halichoerus grypus*. The latter is seldom seen in the North Sea but is widespread along our western coasts from the Scilly Isles and Cornwall and along the shores of Wales and Ireland to the Hebrides, and to Orkney and Shetland in the far north. Like all seals it comes inshore annually to its nursery grounds. Here the adventurous naturalist will find a rich field for observation; but all may read about, and view in photographs of great beauty, the shore habits of the adults and young of Ron Mor, the Great Seal of Gaeldom, in Fraser Darling's *A Naturalist on Rona, Island Years* and *Natural History in the Highlands and Islands*. The harbour seal is smaller, not exceeding six feet, which is only two-thirds the length of a male grey seal. It spends much more time ashore, apart from the breeding season, and frequents estuaries, often penetrating for long distances up rivers. It migrates down the east coast of Scotland in the summer in pursuit of salmon, and may often be seen in numbers basking in the sun on rocks and sand banks.

Whales are truly marine animals in that they breed as well as feed in the sea. They are utterly helpless on land, and quickly die when stranded with the chest crushed beneath the great weight of the body. They range in size from five-foot porpoises to the blue and fin whales, which are the mightiest animals ever evolved. From time to time individuals or whole schools of whales are stranded on our shores. Between 1913 and 1937 such appearances of 819 whales of 20 different species were officially recorded. A large number of these records refer to odd porpoises, an average of some 12 of which are stranded annually, but occasionally they refer to such spectacular and embarrassing events as the casting ashore in 1927 of over 150 false killer whales in Dornoch Firth. The British Museum (Natural History) keeps official records of all whales stranded on British coasts and should be informed without delay when these are seen.

The intertidal plant life is as characteristic as the animal life. But in addition the sea extends its influence further inland over static plants than over mobile animals. Cliffs, sand dunes and salt marshes which fringe the shore are all occupied by a typical flora of halophytic or salt-loving plants which grow only in a salty soil and in a salt-laden atmosphere. Such plants include the marram grass, *Psamma arenaria*, with deeply running stems and roots which bind the loose sand of dunes, the succulent saltwort and the sea aster of salt marshes, and the thrift, sea-lavender and stonecrop of the cliffs. Full consideration of this flora lies outside the scope of this book but one of these plants deserves brief mention, namely the rice grass, *Spartina Townsendii*, which is actively engaged in many coastal areas in the conversion of muddy shore into dry land. It is a coarse, vigorously growing grass, usually from 2 to 2½ feet high, though sometimes much taller, which establishes itself on tidal mud flats with the aid of its long downward and outward growing roots. This plant was first noted in 1870 in Southampton Water and rapidly spread until it covered wide areas. It has since extended east and west along the south coast, to the Isle of Wight and across the Channel to the French coast. More recently it has appeared in other muddy areas such as the upper reaches of the Bristol Channel near Burnham-on-Sea. This plant most probably arose by the crossing in nature of a species of *Spartina* which had previously existed in the area with another species, introduced from America. If this be the case, then it provides an admirable example of what is known as hybrid vigour because the new plant is not only larger and more vigorous of growth than either of its precursors, but also able to live in areas which they cannot colonise.

Above high tide level, where spray sweeps the rocks, patches of grey and yellow denote the growth of lichens. Some lichens extend on to the shore where they mingle with the seaweeds which on many rocky shores are the most abundant and conspicuous of living organisms. These are algae, not flowering plants, and are attached by a hold-fast which is not to be confused with a root although it may branch in a similar manner. The difference resides in its concern solely with attachment and not with nutrition like a root. The body of the alga is known as the frond or thallus and may be many feet long in large species.

There are four groups of these marine algae most easily distinguished by their characteristic, although not altogether invariable, colour, namely the blue-green (Cyanophyceae), green (Chlorophyceae), brown (Phaeophyceae) and red algae (Rhodophyceae). They occur roughly in this order from high tide level down the shore. The blue-greens are most difficult to recognise being fine filamentous plants often forming a slimy scum on the surface of rocks or in pools usually at or above high tide level. The green seaweeds are especially abundant where fresh water runs down the rocks. The brown seaweeds form the great bulk of intertidal vegetation and great masses of fucoid weeds carpet rocky shores, to be replaced about the level of low water of spring tides by the massive tangle weeds. The red seaweeds in general appear lower on the shore but mingle with the brown seaweeds here and in shallow water, but they descend much deeper and are the last plants to disappear as the fading light intensity in increasing depths makes plant life impossible on the sea bottom. Broadly speaking, therefore, it may be said that light intensity controls the distribution of these algae, the greens and browns being comparable to sun plants and the reds to shade plants. The quality of light is also important because sea water acts as a filter permitting much greater penetration of green and blue rays than of red or violet. For this reason red animals and plants appear black to the diver and in the case of the plants this makes them very efficient absorbers of the blue light which alone penetrates to any distance below the surface and then only in clear water.

One genus of flowering plants, the eel-grass, *Zostera*, has taken to life in the sea. It has thin, band-like leaves, inconspicuous flowers and creeping stems and roots that enable it to live on a soft bottom of sand or mud where algae could not attach themselves. It needs some degree of shelter and is usually to be found in protected bays and estuaries. A small species, *Z. nana*, occurs on muddy shores where it is freely exposed but the larger *Z. marina* lives lower down on bottoms of sand or muddy sand and only the upper fringes of its beds are exposed at the lowest ebb tides.

CHAPTER 4

THE BACKGROUND OF THE SEA

"Who can say of a particular sea that it is old? Distilled by the sun, kneaded by the moon, it is renewed in a year, in a day, or in an hour."

THOMAS HARDY : *The Return of the Native*

T H E vast majority of this host of widely differing animals and plants which inhabit the shore have come from the sea. They are far from being uninfluenced by the land, to the margins of which they cling, but they remain essentially marine creatures. Many spend their early life drifting in the surface waters and when they later settle on the shore they continue to obtain food as well as oxygen from the water spread over the shore by the flowing tide. It is therefore impossible to paint an adequate picture of shore life without first sketching the general background of conditions of life in the sea. Considered as a medium for life, the open sea is as constant as the shore—with its alternation of immersion and exposure and its wide extremes of temperature and of salinity—is inconstant. Some realisation of the stability of the marine environment is essential before we can rightly appreciate the magnitude of the problems of existence and of reproduction, which have been faced and solved by the inhabitants of the continually changing environment of the shore.

. Living matter is so complex and so delicately balanced that it can exist only within a very narrow range of temperature, and within a rigidly balanced medium of salts, either outside or, in the case of the inhabitants of fresh waters and the land, within the body. It must also have a source of energy ; for plants the rays of the sun, for animals the organic matter in which the plants have stored this energy. The sea represents the simplest medium for life because, at any rate in shallow, illuminated waters, all these conditions are fulfilled. Life can exist in all temperatures found in the sea, from the maximum of about $35°$ C. in the surface waters of the tropics to the minimum of a little under $-2°$ C. on the margins of the polar ice or in the abyssal depths of the oceans. Around British coasts temperatures range between 6 and $15°$ C. In this vast volume of water, moreover, temperature changes very slowly and its inhabitants can easily adjust themselves to the slow rise in the summer and fall in the winter. The salt content of the sea is remarkably constant and when increased by evaporation, as in the Red Sea, or decreased by the inflow of many rivers, as in the Baltic, this causes no change in the relative proportions of the various elements. It is primarily the balance of salts that matters and in this respect the sea is just as much a balanced medium for life as is the blood within our own veins.

It is because life is so much more easily maintained within the sea than in fresh water or on land that we are reasonably confident that it evolved in the early oceans. From thence it gradually spread on to the shore and into the estuaries and so into the rivers and lakes and on to dry land. For a variety of reasons only a relatively few types of animals have succeeded in the enormously difficult task of adapting themselves for life within the foreign media of fresh water and of the atmosphere. In the latter they evolved further to produce the highest types of life—ubiquitous and infinitely varied insects and spiders and the higher vertebrates which culminate in man. But to this day the sea remains much the richer in the diversity of the types of animals that live within it. It is otherwise with the plants which early established themselves on the surface of the land, where they evolved into the mosses, fungi, ferns and flowering plants which now cover its surface.

Despite its vast extent, some two-thirds of the surface of the globe, and the obscurity of its depths, which average over two miles and descend in places to more than six miles, the sea is hardly anywhere devoid of life. The discovery of the strange creatures which inhabit the depths of the oceans was one of the great achievements of British zoologists in the middle years of the nineteenth century. Led first by Edward Forbes and later by Wyville Thomson, they carried the dredge deeper and deeper. Their labours culminated in the voyage of H.M.S. *Challenger* which returned, after sailing round the world, with a vast wealth of knowledge about the sea and its inhabitants in all latitudes and in nearly all depths. Upon the massive series of scientific reports later issued, the modern science of oceanography has been built. And into this greater framework of knowledge the shore observations of Gosse and his contemporaries and successors have been fitted.

It is as true for the sea as it is for the land that all flesh is grass. As we have already seen, plants must come first because they only can utilise the energy of sunlight to build up organic matter from carbon dioxide and water and simple inorganic salts. Initially they form carbohydrates (i.e. sugars and starches) and fats and then from these, by making use of appropriate nutrient salts containing the necessary nitrogen, phosphorus and sulphur, they further elaborate the more complex proteins. It is on these three types of organic compounds that animals alone can feed, in other words obtain the material and energy needed for maintenance, growth and reproduction. Animals must therefore feed on plants, or on other animals which have in their turn fed on plants. So in this brief survey of life in the sea we must begin with the sun which provides the energy on which all life depends.

The surface of the land is everywhere illuminated by the sun, but the sea has depth as well as surface. The rays of the sun are quickly absorbed as they penetrate the sea, some two-thirds of the incoming energy being absorbed in the first three feet even in the clearest water. At depths varying with the latitude and season of the year all is absorbed so that in the greater part of the ocean depths continuous darkness prevails. The plant life, essential ultimately for the maintenance of life at all depths, is therefore confined to the shallow layer of illuminated water near the surface. Marine vegetation is obvious along rocky shores, but these marginal seaweeds, however dense they may be locally, are of negligible importance in the economy of marine life. What matters is the microscopic plant life of open waters. This is part of the great community of living things that drift passively in the surface waters, and is known as the *plankton*, just as the actively swimming fish, squids and whales constitute the *nekton* and the bottom-living creatures the *benthos*.

The plankton is at once divisible into its plant members, or *phytoplankton*, and the animals or *zooplankton*. The former consists of three different types of excessively minute plants, although collectively they may occur in such vast numbers as to produce the olive-green colour characteristic of polar seas in their brief summer of exuberant life. First there are diatoms, which are encased in delicate skeletons of silica and which are everywhere abundant. With them, but commoner in warm than in cold seas, are peridinians or dinoflagellates, each with two whip-like processes for locomotion and some with delicate skeletal plates ; and finally vast numbers of still more minute organisms, naked and excessively delicate, which are known collectively as the *nannoplankton* and not all of which may be true plants. Although, as individuals, so minute, in bulk the members of the plant plankton represent an enormous mass of vegetable matter. The yield in plant material from an area of shallow fertile sea water is equivalent to that from a similar area on land. The only difference is that in the sea this plant crop is dispersed through some depth of water and is not concentrated on a flat surface like the grass of a meadow. The animals which feed on these plants are of course similarly spread out through the layers of the sea.

We have referred just now to a fertile area of sea water. This implies that the fertility of the sea varies from place to place as does that of the land. There are regions of immense productivity such as those that support the great fisheries around our own coasts and on the Newfoundland Banks, and there are desert areas such as the Sargasso Sea, the deep blue of whose waters indicates the absence of appreciable quantities of suspended life. Deserts on land are associated with lack of rainfall, in the surface waters of the sea with lack of the essential nutrient salts containing the nitrogen and phosphorus needed by plants to form proteins. These nutrients occur in immense quantities throughout the sea but are only available for use by plants when they occur in the sunlit waters near the surface. And by far the greater quantity is locked away in the deep waters remote from light.

It follows from this that only where these nutrients are in large, and constantly replenished, supply in the surface waters will the sea be fertile. In other words a fertile area is one in which there is ready mixing of the waters from top to bottom or where deep currents rise to the surface. If the seas were perfectly still there would be no such fertile areas ; indeed the surface waters would be unable to carry any appreciable amount of life. This is what does largely occur in areas like the Sargasso Sea. The surface waters, warmed by the sun, float as a distinct layer upon the colder and therefore heavier water below them. The sea is divided horizontally into two layers which represent totally different environments for life. The warm and illuminated upper layer is known as the *troposphere* and the cold, dark regions below which extend to the floor of the ocean are called the *stratosphere*. The two are separated by a narrow zone of water in which the temperature falls rapidly ; there is no gradual falling off in temperature from surface to bottom. This zone is the discontinuity layer or *thermocline*. While it persists it effectively prevents any appreciable mixing of the waters above and below it and so bars the upward passage of nutrients. In regions like the Sargasso Sea this barrier to fertility is always present ; only when it is broken down can the immense reserves of nutrients contained in the deeper layers become available for the nourishment of the plant life in the surface waters.

This essential mixing of the waters can come about in two ways. The surface temperatures may be lowered in winter until they are as low as, or lower than, those of the deeper water. Then, particularly when the water is violently agitated in stormy weather, there is free mixing from top to bottom. This happens in temperate seas such as those that surround these islands. Every winter the nutrient salts in the surface waters are replenished so that, with the coming of increased light and warmth, there is a spectacular burst of surface life, known as the spring increase. Then, when the nutrients have been used up, the productivity falls off, although there is a second, smaller increase before all life dies down with the onset of winter.

But the layers of the sea are also mixed by the action of ocean currents. The sea is far from being a stagnant mass of water. If it were so, not only would the surface waters carry little life but depths and bottom would be lifeless because devoid of oxygen. The sea is a great thermodynamic machine receiving heat in the tropics and losing it in the polar seas. In consequence continual movement is set up throughout this great mass of water. Cold water, rich in the oxygen which permits life and also in nutrient salts, flows along the floor of the ocean from the polar regions to the equator. Currents in other water layers flow back. These deep currents may rise to the surface with dramatic effect on fertility. The immense populations of sea-birds along the coasts of Chile and Peru feed on fish which in turn are nourished by the dense plankton maintained by a constant upward movement of deep water rich in nutrients. The excrement of these birds forms the deposits of guano which are the most concentrated form of natural fertiliser.

The fertility of European seas is profoundly affected by the inflow of currents from the Atlantic. These owe their origin to the mighty Gulf Stream which, forced forward by the piled-up waters in the Gulf of Mexico, flows out through the narrow Straits of Florida at a speed of $3\frac{1}{2}$ knots. It loses speed and spreads out widely before, as the now ill-defined Atlantic Drift, it reaches our coasts. It then flows round the north of Scotland into the North Sea and also, to a more variable extent, into the mouth of the English Channel. Upwelling of deep water on the Continental Slope and, in the North Sea, the presence of swirls and eddies which mix the shallow waters from top to bottom are also prime causes of the rich fertility of our surrounding seas. Further, the distribution of marine animals and plants is controlled largely by temperature and this is raised by the warm Atlantic currents.

Given, therefore, light and a sufficiency of heat from the sun, the fertility of the sea is in direct proportion to the amount of nutrients. First come the plants and then the animals which feed upon them. The animals of the plankton are a most varied assemblage including members, at some stage in their life history, of almost all types of animals and ranging in size from large, gently pulsating jellyfish to minute, darting crustaceans and microscopic protozoans little bigger than the plants. All drift passively in the surface waters, such powers of movement as they may possess enabling them to do little more than maintain themselves near the surface although many also possess droplets of oil, bubbles of gas or long spines to prevent them from sinking. The bulk of these creatures feed on the phytoplankton. This is the case, for instance, with the copepod crustaceans already mentioned as amongst the most numerous of all animals. Other animals of the plankton, such as jellyfish and their allies and the slender arrow-worms, are carnivorous and devour fellow members of the zooplankton. In addition surface-living fish, notably herring, pilchards, mackerel and even our largest fish, the twenty-foot basking shark, are plankton feeders. The vast whalebone whales of the Antarctic, up to one hundred feet in length and as many tons in weight, feed exclusively on krill, relatively large planktonic crustaceans rather like small prawns.

During the spring and summer the planktonic population of inshore waters is swollen by the addition of the early free-swimming stages of many animals which spend their adult life on the sea bottom or on the shore. Such " larval " stages, frequently totally unlike the adult, are known as temporary plankton to distinguish them from the permanent members which spend their entire lives floating, mostly in the surface waters, although in winter they may sink deeper. No animals are more typically dwellers on the shore than the acorn-barnacles which grow in a dense bank near high tide mark on every rocky shore. But after they have spawned in the spring the surface waters are soon thick with the minute crustaceans which Vaughan

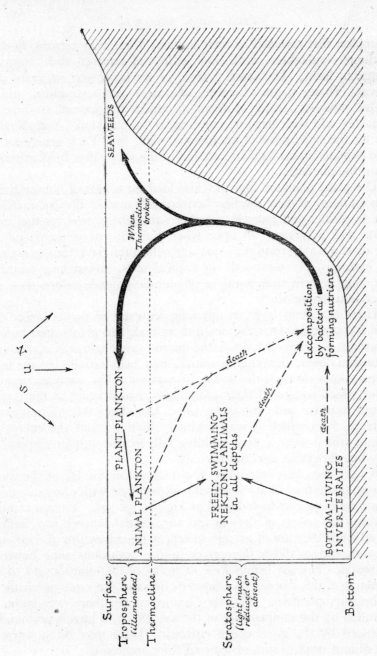

FIG. 21—Simplified diagram indicating cycle of life in the sea. For full explanation see text.

Thompson first revealed as their larvae. There they remain, feeding on the phytoplankton, until they settle and undergo their dramatic change in form. The same is true, as we have seen, of crabs and prawns. Young stages of worms, of starfish and sea-urchins, and of bivalve molluscs, many of them as unlike the parent as those of barnacles, add their teeming numbers to the plankton. And all compete for the same stock of microscopic plant life. Each generation of these animals makes its own invasion of the shore after brief existence within the plankton.

The plant plankton also provides food for a host of bottom-living animals, many of them dwellers between tide marks. Bivalve molluscs, such as cockles and mussels, together with tube-worms, sponges, sea-squirts and polyzoans, all open when the tide covers them and, in their diverse ways, strain the water and collect this food. Sea-anemones and hydroids, like the corals on tropical reefs, extend long tentacles with stinging cells that paralyse planktonic animals before they are passed into the mouth.

There is a final stage in this long sequence of events before the cycle is complete. After the death of animals and plants their bodies, from those of massive whales to the microscopic single cells of diatoms, are broken down. Scavenging animals may help initially but the final process comes through the action of bacteria. The complex organic matter which composed their protoplasm reverts again to the water, carbon dioxide and inorganic salts, including the all-important nutrients, from which it was originally built up and the energy of sunlight which went to that building falls to the ultimate disposal of lowly but altogether essential bacteria.

We see then how impossible it is to consider the life of the shore without realisation of its continual interactions with the sea which I have endeavoured to indicate in Fig. 21 (p. 51). To many shore creatures the waters of the sea are an essential nursery and feeding ground until they are of an age to repeat the adventure of their first ancestors and establish themselves in the hazardous zone between tide marks. The sea brings food to many of the animals and to all the plants of the shore ; the seaweeds demand the same nutrients as do the phytoplankton. The very nature of the shore population is controlled by the temperature of the sea, which in turn is profoundly influenced by the great ocean currents. These may bring warmth from distant tropical suns or icy cold from arctic seas.

CHAPTER 5

THE MOULDING AND NATURE
OF THE SHORE

"The insular peaks and blocks receive the incoming surge in an overwhelming flood, which, immediately, as the spent wave recedes, pours off through the interstices in a hundred beautiful jets and cascades ; while in the narrow straits and passages the rushing sea boils and whirls about in curling sheets of snowy whiteness, curdling the surface ; or, where it breaks away, of the most delicate pea-green hue, the tint produced by the bubbles seen through the water as they crowd to the air from the depths where they were formed—the evidence of the unseen combat fiercely raging between earth and sea far below."

PHILIP HENRY GOSSE : *A Year at the Shore*. 1865

W H E R E the great background medium of the sea strikes the land there comes into being the narrow belt known as the shore, upon which so many of the sea's one-time inhabitants have taken up permanent or temporary residence. These shores, of firmly stretching sand, of jutting rocks, or of wide wastes of mud, are all the product of long interaction between sea and land. The results of this conflict of opposing forces depend in equal measure on the nature of the land and on the direction and force of the waves and tidal movements of the sea. The eventual balance achieved, in some cases relatively enduring and in others poised in unstable equilibrium, is of supreme importance to the animals and plants. Each type of shore presents a different series of problems for solution by the innate adaptability of living things. The fascination of the shore to the naturalist is immensely enhanced by the rich variety of environments presented to his inquiring gaze.

The full story of shore formation is a large and complicated one to be told adequately only by an experienced geographer. Yet this account of shore life would be inadequate without some mention of the manner in which the different types of shore have come into being around our coasts.

The striking contrast between the sharp irregularity of rocky shores on the one hand and the graceful curves of sand beaches and the wide expanses of mud flats on the other represents the outcome, respectively, of the destructive and constructive forces of the sea. What the sea wrests from the land at one place it returns, usually in more extensive area (though not volume), elsewhere. On balance, therefore, the area of these islands is increasing. From the standpoint of the shore inhabitants—our sole justification for considering the matter—they are being pushed forward in certain areas and being withdrawn seaward in others. Waves are the prime cause of destruction, although within restricted channels such as the Menai Straits and in tapering estuaries like the upper reaches of the Bristol Channel and the Solway Firth, the scouring of tidal currents plays a part. The force of a wave when it strikes the shore depends less on the strength of the wind than on the " fetch " or extent of open water over which this has blown. It is for this reason that the breakers along the north coast of Cornwall, where the fetch of the wind is the wide Atlantic, are so impressive. The effective action of the waves is also influenced by the depth of the water offshore. If deep water extends to the base of a cliff, the waves move up and down the cliff face with little destructive action. But if the water is shallow the circular movement of individual particles of water, which causes the rollers of the open sea, is translated into vertical and finally into forward movement when each wave in turn crashes over into a breaker. Then a tremendous force, up to 25 tons to the square yard, may be released against the shore. Massive breakwaters at Port Erin and at Wick, and at many other places, representing many thousands of tons, have been thrown aside by such waves.

The destructive action of waves that break along the base of cliffs is greatly enhanced by the stones their breaking waters pick up and hurl against the face of the rock. Moreover air is caught and compressed within cracks and assists in erosion especially at the moment of its sudden expansion when the pressure of water is released. In the past ice played an important role. Much also depends on the nature of the cliff. Soft rocks are worn away quicker than hard ones while the thickness and the angle of slope of the strata are important. Cliffs composed of alternate bands of hard and soft rock are relatively quickly eroded especially if the strata dip seawards when the hard bands will be undercut and collapse. The animals and plants are not without influence between tide marks. Certain animals, with which we shall make acquaintance later, bore into rocks. Seaweeds, on the other hand, protect the rocks, acting as a blanket against the force of the waves and insulating them from the effect of extreme cold.

Aided in these divers ways, the sea cuts into the base of cliffs the walls of which eventually come crashing down. The mound of debris delays further destruction until it has been cleared away, but when this has been done the sea advances afresh over a broad wave-cut bench of cleanly swept rock such as that so well displayed along the coast of Yorkshire southward from Hunt Cliff. Where the strata alternate in hardness with a lift seaward, the successive ridges of harder rock form the scaurs with intervening rock pools described by Leo Walmsley in his stories and reminiscences of " Bramblewick " (Robin Hood's Bay).

The material so broken away from the land may long persist as boulders or, where the waves are powerful or the rocks are soft, be quickly ground down to rounded pebbles and shingle and so to sand and finally to the finer particles of mud. The more finely divided and lighter these particles become the further can they be transported by the sea. Such transport is largely the result of wave action. Only in very sheltered waters, at the head of deep and narrow bays, do waves normally hit the shore at right angles, so producing a purely up-and-down movement of sand on the shore. On open shores the waves usually come in somewhat obliquely, depending on the direction of the prevalent winds. Consequently particles are driven up the shore at that angle but are then carried almost directly down the slope of the beach in the backwash. The resultant beach drift thus takes the form of a zigzag movement which is shown diagrammatically in Fig. 20 in L. Dudley Stamp's *Britain's Structure and Scenery* in this series. The accumulation of shingle against projecting groynes and a breakwater at Folkestone due to this coastwise drift is illustrated in Plate IX in that book. The ultimate effect of this movement is to sort out the eroded material according to the size of the particles. Thus shingle beaches are found nearest the site of marine erosion, sand is carried farther away and mud farther still. In its lateral carriage the finer material is deposited in the shelter of inlets or wherever wave action is obstructed, by the outflow of rivers, or by conflicting tidal and wave movements when long spits may be formed.

Sandy beaches are built up in bays which range in size from narrow inlets between adjacent headlands of rock as on the north Cornish coast to wider expanses along the Lancashire coast or in Carmarthen Bay. Such a beach rises gently inshore up to about the level of mean high tide when the curve increases, the contour of the beach resembling, as Shaler points out in *Sea and Land*, the half of a catenary curve such as is made when a rope hangs free between two points of attachment. The curvature varies continually with the force and nature of the waves; some are constructive, carrying more material upshore in the " swash " than downshore in the backwash, others are destructive with the greater effect in the backwash. There is also a cycle of changes associated with that of the tides, from springs to neaps and back to springs again. The curvature at any one time depends on the balance between the upward carriage of sand in the swash and its downward movement in the backwash.

Sand is a very beautiful medium. Every child realises this from its first moment on the beach. But the wonder grows no less with the considered judgment of maturity. Sand is vastly more durable than the massive pebbles which are quickly ground down against one another by the constant pounding of the waves. Although pebbles vary greatly according to the nature of the rock of which they are composed, the durability of many of them is estimated at little more than a year. But a grain of sand must be of durable texture to have withstood transport without being further reduced to a particle of mud. Each grain is so small and with so little space between it and its fellows that water is held around it by capillary action. Hence no two grains touch one another and the sand is not ground to fine powder even when powerful waves, coming in at a rate of some five every minute, pound with the force of hundreds of tons along a stretch of beach.

Pebbles roll over each other and so would crush animals or plants that attempted to live among them ; a pebble beach is the most barren region of the shore. But unless churned with exceptional force, as on the more exposed Atlantic beaches, the water-laden sand is a wonderful medium for living things. It gives protection when the tide is out and a secure substratum on which an animal may move or in which it may burrow when covered by the sea. The dry sand about high tide level or on sand dunes is barren. Here the grains rub one against the other and in movement may give out the sharp " singing " note heard by travellers over certain sands. The enduring wet sands of the sea shore are protectors of the land and over their firm expanse the waves break with fruitless fury.

Muddy shores are typical of estuaries and of the inner areas of long inlets like many of the shallow sea lochs along the west coast of Scotland. Especially where they are consolidated by plant growth, notably by *Spartina Townsendii*, such shores represent areas of potential new land. For a variety of reasons to be discussed later, they present greater problems for effective colonisation than do either rocky or sandy shores and carry a correspondingly more restricted and specialised population.

Between each major type of shore, of rock, pebble, sand or mud, there is often intergrading. Both composition and configuration of rock influence the population. Volcanic rocks tend to weather smooth so that attachment to them is difficult but the fissures and crannies of eroded limestone and still more the deep crevices between the ragged ridges of weathered shales and slates harbour a dense population. The character of the rock surface, finely or coarsely grained, hard or soft, is important, and so is the direction and amount of slope. Seaweeds in general are less abundant on limestone than on other rocks. Sand and mud also vary in consistency, the most important factor being the amount of contained organic debris. Where this is abundant, bacterial action causes the decomposition responsible for the blackening of the under layers of sand and the strong smell of sulphuretted hydrogen.

The area covered and uncovered by the tides varies greatly in width according to the nature of the shore and the extent of the tides. In general shingle beaches and boulder-laden shores are steepest while muddy and sandy shores have the most gentle slope. Wide stretches of sand are exposed at low tide in such regions as Holderness and along the coast of Lancashire, and of mud in the Bristol Channel and mouth of the Thames. The range of tidal movement varies greatly. In the open ocean this is no more than three feet, but as the speed of the great tidal waves which travel round the globe is checked by shallowing seas and then by coasts, their height increases. The average range of tides around the British Isles is 15 feet at springs, when there is a united pull of sun and moon on the water, and of $11\frac{1}{2}$ feet at neaps when these forces of attraction are at right angles to one another. But the configuration of the land also has an important effect. Tidal range is increased in funnel-shaped inlets, notably in the Bristol Channel, where the tidal range at Chepstow is some 42 feet at springs and 21 feet at neaps. Conflicting water movements, on the other hand, may reduce the maximum tidal range—e.g. to about 5 feet between Swanage and Portland although it is 38 feet across the English Channel in the Bay of St. Malo.

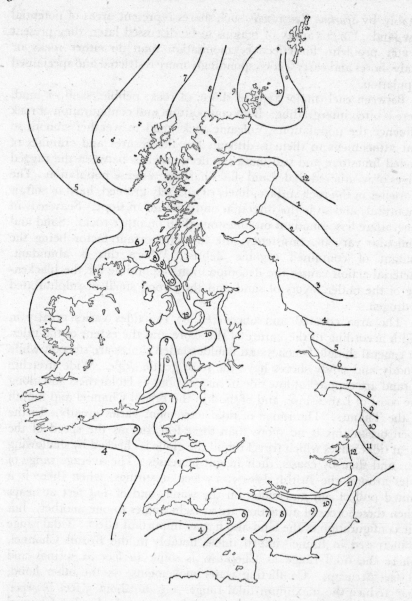

FIG. 22—Co-tidal lines, showing the course of the tidal wave in British Seas. (Data from Flattely and Walton, *The Biology of the Sea-Shore*.)

Two tidal waves of about the same height strike our coasts during the lunar day of 24 hours and 50 minutes. Approaching from the west they are divided by the projecting mass of Ireland. The southern branch is again divided when it strikes the peninsula of Cornwall, one part passing into the English Channel and the other into the Bristol Channel and northward into the Irish Sea. The northern branch, after sending a minor wave through the North Channel into the Irish Sea, travels round the north of Scotland and down the east coast into the North Sea. It takes 12 hours to reach the southern end of this, where it blends with the succeeding wave which has passed up the English Channel as shown in the map giving the co-tidal lines around the British Isles (Fig. 22, p. 58). The times of high and low tide on any day thus vary widely in different areas around our coasts. To the shore collector this is important in connection with the incidence of spring tides at the ebb of which the greatest area of shore is exposed for his searching. These tides come twice in each lunar month with the maximum range some two days after new and full moon and " good " tides over some five days around each peak. The neaps come in between, after the first and third quarters of the moon. There is also a seasonal effect, the greatest spring tides occurring at the March and September equinoxes. The wind has also to be considered, especially in enclosed areas where a powerful onshore wind will bank up the water and prevent a good ebb while an offshore wind will have the opposite effect.

The shore, intertidal or littoral zone, may be defined as that region which is bounded on its landward side by the extreme high water level of spring tides (EHWS) and on its seaward side by the extreme low water level of the same tides (ELWS). But true shore animals and plants are to be found living above extreme high water level on what is technically dry land but is drenched by the splash and spray from breaking waves at high tide. Although the term has been variously used, this region will be called the splash zone. It is environmentally a part of the shore with its own particular fauna and flora. In countries where steady onshore winds cause a great surf this area may be further sub-divided into wash, splash and spray zones but such conditions hardly prevail here.

On the shore itself we recognise a series of levels between the extremes already mentioned. Passing down the shore there is first the level of mean high water of spring tides (MHWS), followed by that

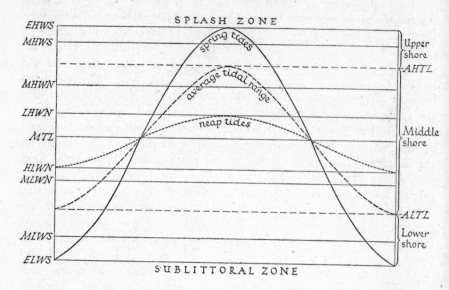

FIG. 23—Shore levels in relation to tidal changes. For full explanation see text.

of mean high water of neap tides (MHWN), then lowest high water of neap tides (LHWN), next comes mean tide level (MTL), then highest low water of neap tides (HLWN), followed by mean low water of neap tides (MLWN) and finally mean low water of spring tides (MLWS). These levels, denoted by their customary abbreviations, are shown in Fig. 23.

The extent to which the tides range up and down the shore is also shown in the figure, from the extremes of springs to those of neaps. But clearly neither represents normal conditions; these are best indicated by the average tidal range, which is shown between the other two. On this basis we can erect two new horizontal boundaries, shown by broken lines in the figure and representing the upper and lower limits of average tidal movement. Although this is a new departure, I am now going to divide the shore into three zones, the upper shore above the average high tide level (AHTL), the middle shore between this and the average low tide level (ALTL), and the lower shore between this and the extreme low water of spring tides. The upper shore is thus a region permanently uncovered except when the tides exceed average range, which is also the only time when the lower shore is bared. The middle shore, on the other hand, experiences in the main daily submergence and uncovering. This subdivision of the shore does appear best to correspond to our present knowledge of the vertical distribution of the shore population, while it has the merit of much greater simplicity than many previous schemes of subdivision.

Below the shore stretches the sub-littoral zone which extends, with gradually increasing depth, to the margin of the continental shelf at a depth of some 100 fathoms (600 feet). Here the slope of the sea bottom becomes steeper and descends down the continental slope to the great depths of the ocean floor.

On the restricted area of the intertidal zone and under the widely differing conditions of rock, sand or mud created by the interactions of sea and land, there dwell the many types of animals and plants which were briefly surveyed in Chapter 3. It is to the consideration of the distribution and habits of the commoner of these that we may now suitably pass. We shall encounter many ingenious adaptations of structure and habit. But certain characteristics all, in varying degree, must possess, namely the functional capacity to withstand the effects of a wide range of, and sudden changes in, temperature and salinity. Added to this, in many instances, is the power to resist or circumvent in one manner or another the effect of exposure to the air and of mechanical pounding by the waves. It is the possession of these capacities which represents the fundamental difference between the inhabitants of the shore and those of the sub-littoral zone.

CHAPTER 6

THE SEAWEEDS OF A ROCKY SHORE

"How softly the feathery sea-groves are waving !
    Their plume-tufts of purple, and scarlet, and green,
The pure and clear element gently is laving :—
    While tiny swarms merrily sport them between.".
                    PHILIP HENRY GOSSE : *The Aquarium*. 1854

W E can most suitably begin our survey of shore life on the rocks. Except in areas exposed to exceptionally heavy seas, such as those that roll in from the Atlantic on the north coast of Cornwall, the first thing to catch the eye when the tide is out will be the rich growth of weeds clothing the rocks from the splash zone down to the lowest tidal limit and beyond. For a variety of reasons it is fitting that we should start by considering these seaweeds. Plants form the ultimate source of food for all animals. Were we dealing with a restricted area on land, or better still an island, we could say that upon the vegetation there growing all the animals, apart from sea birds, were dependent. But this, as we have already seen, is not true of the shore. The sea-weeds do provide food for a variety of small animals that browse upon them, their torn fragments enrich the food content of bottom material taken in by worms and scavenged over by many crustaceans, while after they are cast up to dry and decay on the strand line they are at once the home and feeding ground of many small crustaceans, insects and worms. But many shore animals, and in turn the animals that may prey upon them, are nourished by the never-ending supplies of plankton brought in with the advancing tide.

It remains true, however, that many animals that do not feed on seaweeds are dependent on them for shelter, alike from the force of the waves when the tide is in and from the heat of the sun or the drying action of the wind when the tide is out. Remove the seaweeds and a great proportion of the animal life would be either swept away or die from exposure. Their widely spreading fronds provide attachment for hydroids, colonial sea-squirts, sea-mats and the like, which find their food in the plankton. For all these reasons, as a source of food, as a protective blanket against wave action and desiccation, and

63

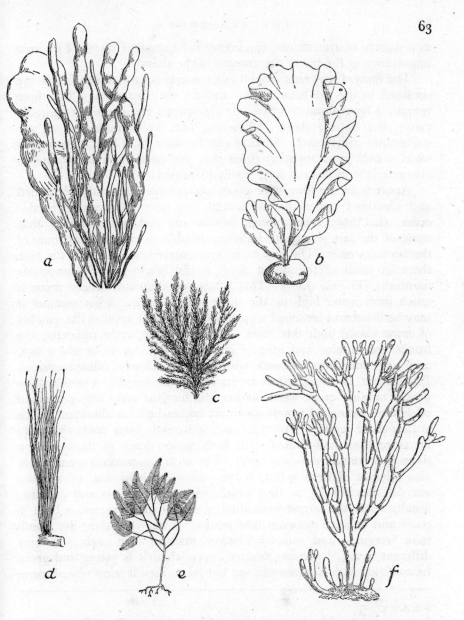

FIG. 24—Common Green Seaweeds (Chlorophyceae), variously reduced. *a. Entero-morpha intestinalis ; b. Ulva lactuca ; c. Cladophora rupestris ; d. Chaetomorpha aerea ; e. Bryopsis plumosa ; f. Codium tomentosum.* (From Newton, *Handbook of the British Seaweeds*, British Museum (Nat. Hist.), 1931.)

as a surface of attachment, the seaweeds represent a factor of the first importance in the life of the animals of the shore.

The flora of seaweeds around our coasts is large ; some of these are confined to the sub-littoral zone and do not concern us ; but there remain a large number which, in widespread abundance or sporadic rarity, live on the shore. Moreover, like the animals, not all are universally distributed. All that can be done in a book covering so wide a field is to mention those that are most common or are so characteristic in form as to be easily identified.

Apart from the filmy blue-green algae, which can only be studied and identified under the microscope, we have to consider, in this order, the three types, green, brown and red, remembering that some of the last group are indistinguishable in colour from some of the brown weeds. The first to be encountered, often above the true shore in pools of the splash zone, is the bright-green *Enteromorpha intestinalis* (Pl. 2a, p. 5). This is always abundant where there is much fresh water high on the shore where, during a hot summer it may be dried and bleached a pure white so that it appears like patches of snow above high tide level. As the specific name indicates, the fronds are tubular and long (Fig. 24a, p. 63), up to nearly 2 feet, and are sometimes inflated with gas. A closely related species, *E. compressa*, is distinguished by its branching fronds. On the shore proper another green weed, darker and forming erect tufts composed of slender, branched fronds up to six inches high, is abundant. This is *Cladophora rupestris* (Fig. 24c) and it extends from pools that may be more than half-diluted with fresh water down to those low on shore that contain pure sea water. The most conspicuous green weed, also tolerant of fresh water, is the common sea-lettuce, *Ulva lactuca* var. *latissima* (Pl. 34, p. 169) which, with allied species and varieties, usually forms the largest individual masses of such vegetation both in pools and exposed between tide marks. It has very thin, flat fronds with irregular and somewhat wavy margins (Fig. 24b). A very different weed, *Aspercoccus fistulosus*, may when it is young and green be confused with *Enteromorpha* but not later when it turns olive-brown.

PLATE 5
LAMINARIA ZONE exposed at low water of spring tides. Tangle-weed consists of *Laminaria digitata* with a little *Fucus serratus* and red *Rhodymenina palmata* in the background

PLATE 5

PLATE 6

The soft tubular fronds, which range in length from a few inches to 2 feet, grow in tufts in shallow pools and exposed on stones. Each frond tapers to the base, which is attached by a small disc.

Smaller green weeds, largely confined to pools, include the " hog's bristle," *Chaetomorpha melagonium*, which consists of clusters of stiff, dark-green threads up to 5 inches long, and sometimes *C. aerea* (Fig. 24d, p. 63) the finer threads of which may form tangled masses near high tide mark. The most beautiful of them all is the delicate *Bryopsis plumosa* (Fig. 24e) which may be seen growing on the sides of deep pools. The main stems carry fine lateral branches so that, as it is only a few inches high, each plant resembles a bunch of small green feathers. The last green weed we need mention, only to be found occasionally in pools low on the shore although it is often abundant off-shore, is *Codium tomentosum* (Fig. 24f). This is a most handsome plant, consisting of numerous dark-green and rounded branches, sponge-like in texture, which form relatively massive growths sometimes over a foot high.

The brown weeds are much more numerous and in every way more important members of the shore population. They may here be conveniently divided into those that live on the exposed surfaces of rocks and those that are usually confined to the standing water of rock pools. The former consist of the sea-wracks, a group of leathery olive-green weeds all belonging to one family, the Fucaceae. They attain the necessary firm attachment by means of a disc-shaped holdfast and after vainly attempting to break this off by pulling on the weed we realise why these species are able to withstand the continuous pounding and surging of the seas. Each plant has a short stalk and then the frond divides evenly into many branches. When they are sexually ripe the swollen tips of many of the branches are covered with small spots, the conceptacles in which the reproductive cells are produced and from which they issue for subsequent fertilisation. The highest levels of the shore, varying in width according to the slope, are occupied by the channelled wrack, *Pelvetia canaliculata*

PLATE 6
TINY ROCK POOL on upper shore, lined with pink *Lithophyllum incrustans* and containing Beadlet-anemones, *Actinia equina*, Common Limpets, *Patella vulgata*, Flat Top-shells, *Gibbula umbilicalis*, with tufts of *Corallina officinalis* and a young *Fucus* plant. Acorn-barnacles have settled in the hollows of the rock face above the pool

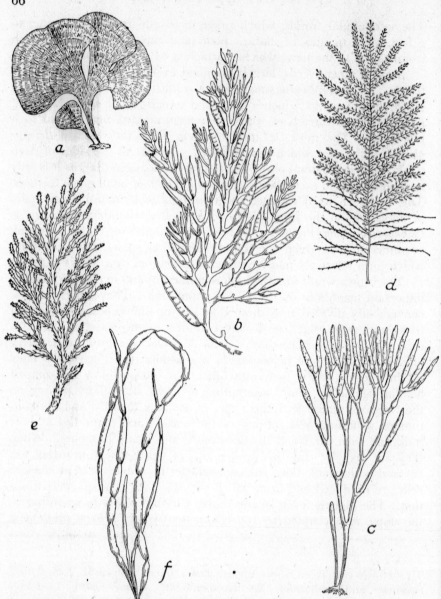

FIG. 25—Common Brown Seaweeds (Phaeophyceae), variously reduced. *a. Padina pavonia*; *b. Halidrys siliquosa*; *c. Bifurcaria tuberculata*; *d. Desmarestia aculeata*; *e. Cystoseira ericoides*; *f. Scytosiphon lomentarius*;

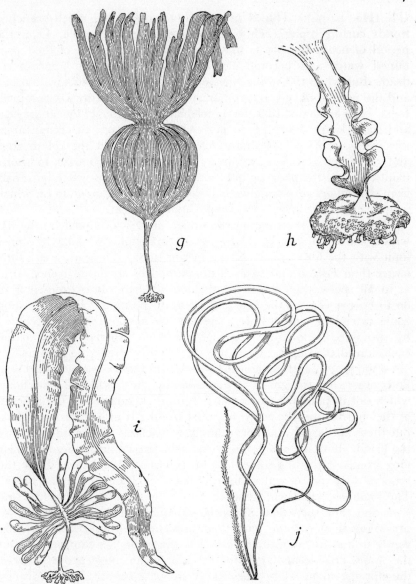

g, Laminaria Cloustoni (showing new growth); *h. Sacchorhiza bulbosa*, showing only holdfast and stem with frilled margins; *i. Alaria esculenta*; *j. Chorda filum.* (From Newton.)

(Pl. IIIa, p. 32). This is unmistakable, the short, much-branched fronds curling along each margin to enclose a channel. Over the period of neaps the plants may lose up to 65 per cent of their contained water and become dry and blackened, to all appearances dead. But when the spring tides extend over them water is absorbed and the normal olive-green colour and the softer texture are regained. It has been estimated that the upper plants are exposed for 90 per cent and those lower down for 70 per cent of the year. In certain areas this weed occurs as an almost moss-like growth in the splash zone and there mingles with a variety of coastal, as opposed to shore, plants. It must experience occasional wetting by sea water but apart from this makes surprisingly small demands on the medium on which it must originally have been completely dependent.

A second narrow zone on the upper shore is occupied by the flat wrack, *Fucus spiralis* (Pl. IIIb), which may mingle along its upper limit with the lower plants of the *Pelvetia* zone. This wrack is a little larger than *Pelvetia* and easily distinguished by its flatter fronds with, as in all species of *Fucus*, a conspicuous midrib. It is exposed from 80 to 60 per cent of the year according to position. The succeeding zone, extending over the wide area of the middle shore, is occupied by two weeds, sometimes together but more usually by one to the exclusion of the other. These are the bladder wrack, *Fucus vesiculosus* (Pl. IVa, p. 33), and the knotted wrack, *Ascophyllum nodosum* (Pl. IVb). Both are easily recognised, the former by the bladder-like swellings which dot in pairs the surface of the fronds, the latter by the absence of the midrib and the more rounded frond. It also carries bladders, but these are larger and borne singly as swellings of the full width of the frond. Both are much larger than the preceding species and their long fronds, extending sometimes to six feet in *Ascophyllum*, drape the rocks in thick masses and so furnish ideal shelter for the animals below. The bladders, filled with gas, serve to buoy the plants when the tide is in and they then float upwards, streaming in the direction of the prevailing tidal current, with their branches well apart and swaying gently to and fro. On the rocks below the anemones can expand, and the snails, crustaceans and other creatures can move about between the upright stems as between the trees of a forest. The buoyancy conferred by the bladders is so great that relatively large stones on which they have grown can be transported for considerable distances.

PLATE V

D. P. Wilson

a. Serrated Wrack, *Fucus serratus* ($\times \frac{1}{3}$)

D. P. Wilson

b. Thong Weed, *Himanthalea lorea*, button-like vegetative fronds with young reproductive thongs growing out of many of them ($\times \frac{2}{3}$)

PLATE VI

Boulders at Millport, I. of Cumbrae, covered by densely growing *Gigartina stellata* with long reproductive straps of Thong Weed, *Himanthalea lorea*

R. B. Pike

The reasons why these two bladdered wracks seldom occur together, and when they do so are sometimes mingled and sometimes in distinct zones, remain to be adequately determined. One thing is clear, namely that *Ascophyllum* is a plant of sheltered waters. It cannot withstand violent wave action, for which indeed the nature of its longer and thinner fronds do not fit it. *F. vesiculosus* has compact growth and stouter fronds with the strengthening midrib characteristic of the genus. But we shall return to this subject later when discussing the general question of zonation.

The lower zone on the shore is occupied by the serrated or toothed wrack, *Fucus serratus* (Pl. Va, p. 68). This has flat fronds like those of the flat wrack but larger and with serrated margins. The absence of bladders separates it at a glance from either of the two preceding species with which it may mingle for some distance above low water line of neaps. It forms an equally dense covering.

In restricted areas two other species may be found. Where much fresh water descends over the shore, especially in land-locked bays, another wrack, *F. ceranoides*, may occur to the exclusion of all others. This has no bladders and most closely resembles flat wrack but is larger, up to 2 feet high, and with a wavy margin to the fronds. On the west of Scotland and in Ireland at the heads of deep lochs on shores of mud and gravel there may be found rounded masses of *Ascophyllum Mackaii*, which is unknown except in its particular habitat of exceptionally sheltered water where it lives unattached. Recent investigations, however, suggest that this may originate from normal attached plants of *A. nodosum* giving rise to a variety of *A. nodosum* var. *Mackaii*.

The most easily identified brown weed on the shore is frequently found near low tide level. Early growth consists of a rounded button-shaped object which then enlarges like a small mushroom and later hollows out with a concave upper surface. Finally from the middle of the depression a bud appears and grows out into a greatly elongated branched strap. This is the thong-weed, *Himanthalia lorea* (Pls. Vb and VI). The small rounded basal region is the frond, the thong which grows from this during the winter bears the conceptacles which ripen in the following summer. One of the commonest of tropical sea-weeds, found in wide abundance on the surface of coral reefs, has a solitary representative on our southern shores in the beautiful peacock's tail, *Padina pavonia* (Fig. 25a, pp. 66-67). From a narrow base this grows out as a thin fan with inwardly curled margins. The surface is marked by concentric lines of iridescent hairs and of white patches of lime. Although technically a brown weed it ranges in colour from yellowish olive over the main surface to reddish near the base and green towards the margins.

Before discussing the nature of the larger, essentially sub-littoral, weeds that are in part exposed by the lowest ebb of spring tides, we should first mention the commoner brown weeds which usually occur only in rock pools. In the deeper pools near the seaward limit of the shore there may be found bushy clusters of the pod-weed, *Halidrys siliquosa* (Fig. 25b, pp. 66-67), like little shrubs sometimes 2 to 3 feet high. The fronds are elongate and compressed, with many short branches bearing long " pods " with pointed tips. These give the weed its common name and make it easy to identify. Each is ribbed transversely, so giving a superficial resemblance to the seed-pods of the gorse, but they are really bladders subdivided into a series of chambers. *Halidrys* is essentially a northern weed which flourishes in arctic seas but a somewhat similar, although rather smaller species, *Bifurcaria tuberculata* (Fig. 25c), is not uncommon in pools along the south coast and is shown in Plate 2b (p. 5). It has the same type of erect growth but with dichotomous branching and occasional simple rounded bladders situated at some distance from the tips of the branches which are swollen with numerous reproductive conceptacles. A more delicate and fern-like growth is characteristic of the species of *Desmarestia* (Fig. 25d) which also grow on the rocky bottoms of tidal pools low on the shore.

Those who have searched pools along the coast of Cornwall or on other southern shores may well have admired the iridescent beauty of what has been called the rainbow bladder-weed, *Cystoseira ericoides* (Pl. 3, p. 12). Small bladders are born along the numerous short branches, the ends of which bristle with spiny processes (Fig. 25e). As in all iridescent weeds, the beauty largely disappears when the plant is taken out of water. Often with it, as shown in Plate 3, is the larger and still more branched *C. fibrosa*. Both are of a yellowish-orange colour and tough in texture and their dense growth renders them admirable homes for many small animals. A weed of much wider distribution and shown in the same plate is *Scytosiphon lomentarius*, which has unbranched fronds, a foot or more in length, usually easily recognised because of the frequent constrictions along their length (Fig. 25f).

The characteristic weeds of the lowest levels, exposed only at the ebb of the greater spring tides, are massive tangles or oar-weeds which here replace the intertidal Fucaceae. They are true plants of the sub-littoral descending to depths of about 15 fathoms before diminished light makes life impossible. They exhibit some degree of zonation at the margin of the shore. The first to be exposed is *Laminaria digitata* (Pl. 5, p. 64) which has a stout, smooth stem secured to the rocks by a branching holdfast (Fig. 27, p. 75) and spreading out above into a wide digitate frond, tough in texture but smooth to the touch. The allied *L. Cloustoni* (Fig. 25g, pp. 66-67) is distinguished by its rugose stem. Moreover whereas *L. digitata*, the better to resist the effects of desiccation, lies prostrate when the tide ebbs, the latter species which occupies a deeper zone remains erect. *L. Cloustoni* grows rapidly, an area denuded by Dr. J. A. Kitching off the coast of Argyll being covered with new plants one metre high in twelve months. At some-what lower levels and more seldom exposed comes *L. saccharina* which has shorter and thinner stems and an undivided frilled frond which floats out for distances of up to 20 feet. It is also more short lived than the others, seldom surviving for more than three years. It owes its specific name to the presence of the sugar, mannitol, which it stores and which appears as a white powder on the surface after the plant has been dried.

Growth in these tangle-weeds occurs not from the tip, as in land plants, but from a growing point just above the top of the stem. From this region a new frond grows out annually and pushes out the old one which is eventually cast off. There is an obvious advantage in such a procedure because the free end is continually being rubbed, or in rough weather pounded, against the rocks by the waves. New growth from here would be difficult and sometimes impossible.

It is not uncommon to find thrown up on the shore large bulbous masses of weed covered with warty excrescences. They are the un-mistakable holdfasts of *Sacchorhiza bulbosa* (Fig. 25h) which may be a foot in diameter and, being hollow and perforated, form a secure retreat for many small animals. The plant has a broad flat stem, spirally twisted above the swollen holdfast and with wavy margins. The frond is divided up like that of *L. digitata* but into more numerous strips. The last of these large weeds is *Alaria esculenta*, or murlins, which has the undivided frond of *L. saccharina* but with a pronounced midrib (Fig. 25i). The fully grown stem frequently bears small, rib-less offshoots. These great tangle-weeds are all of them essentially

northern plants and grow much larger and in greater profusion off
the west of Scotland and around Orkney and Shetland than they do
off the coast of England.

There is a weed, allied to the tangles, which occurs most usually
on shallow sandy bottoms but which can most suitably be mentioned
here. The sea-lace, *Chorda filum* (Fig. 25j, pp. 66-67), is unmistakable,
the undivided fronds being no more than a quarter of an inch in
diameter but extending for lengths of up to 9 yards and often growing
in large numbers together, with the holdfasts attached to scattered
stones or rock under the sand. The surface is very slimy and, although
so slender, the fronds are exceptionally tough and almost impossible to
break. It grows in the basal regions, with the free end continually
dying off. And finally, among the brown weeds, there is a great
assemblage of fine hair-like species of *Ectocarpus* which live largely on
the larger weeds, some merely growing harmlessly upon them, others
being truly parasitic.

The red weeds are less in bulk than the brown, but much richer
in species. Rigorous selection must therefore be practised in their
description, and attention confined to the commoner and more easily
identified, starting with those usually found exposed on the surface of
rocks or of other weeds. Flat rocks in the lower zone of the shore may
be covered with a thick carpet, some 6 inches thick, of the well-known
carrageen or Irish moss, *Chondrus crispus* (Fig. 26a, p. 73). It consists
of much-branched and very tough fronds firmly secured by a flat
rounded holdfast. Although a red weed its colour is variable, frequently
a dull purple but ranging into green and yellow. When covered with
water (and it is not uncommon in pools), it has a beautiful iridescent
sheen. A very similar weed, *Gigartina stellata* (Fig. 26b), often mingled
with *Chondrus*, may easily be confused with it. The only obvious
difference is the somewhat greater terminal branching of the fronds in
Gigartina. This weed also grows alone and in dense profusion, notably
along the western shores of Scotland (Pl. VI, p. 69).

Examination of *Ascophyllum* seldom fails to reveal the deep red
tufts of the delicate filamentous *Polysiphonia fastigiata* growing in
scattered patches upon it. Like some species of *Ectocarpus*, it is an
epiphyte, not a parasite. Another easily identified weed is the purple
laver, *Porphyra* (Fig. 26c), which has the same type of flat-lobed fronds
as the green *Ulva* but is coloured purplish-red or brown. There are
several commonly distributed species of this weed with a wide vertical

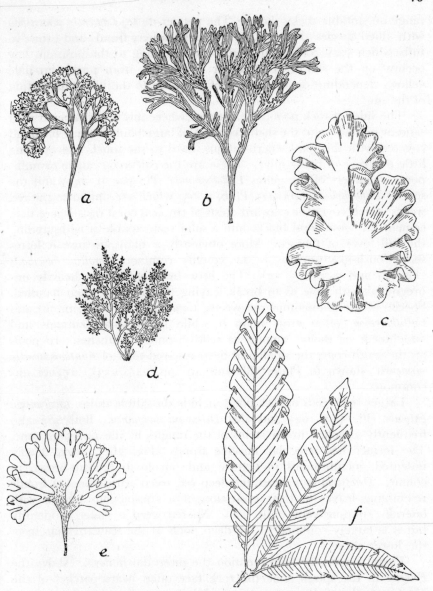

FIG. 26—Common Red Seaweeds (Rhodophyceae), variously reduced. *a. Chondrus crispus ; b. Gigartina stellata ; c. Porphyra umbilicalis ; d. Laurencia obtusa ; e. Rhodymenia Palmetta* (smaller than the allied *R. palmata*) *; f. Delesseria sanguinea.* (From Newton.)

range on suitable rocky shores. The pepper-dulse, *Laurencia pinnatifida* with allied species (Fig. 26d, p. 73), has feathery fronds and grows in tufts which vary considerably in form according to the position they occupy on the shore. The colour also varies from purple to pale yellow, depending on the degree of exposure to the bleaching action of the sun.

The sides of rock pools, the bottom (where this is not covered with sand or pebbles) and the sheltered faces of large boulders may be found covered with thin pink encrustations, hard to the touch and showing little evidence of being alive. These are the calcareous algae or nulli-pores, members of the genus *Lithothamnion* (Pl. 25a, p. 144) and the species *Lithophyllum incrustans* (Pl. 6, p. 65) which are similar to the red weeds which cover the seaward crests of tropical coral reefs where they cement the loose coral blocks into a solid reef capable of withstanding the full force of the sea. More obviously a plant because it forms short, bush-like growths is the equally common *Corallina officinalis* (Pls. 2b and 3, pp. 5, 12). The little branches are so heavily im-pregnated with lime as to break leaving sharp edges when handled. Species of the filamentous red weeds, *Ceramium* (Pl. VIIa, p. 76) and *Callithamnion*, often grow upon it while many small animals find attachment or shelter among its reddish-purple branches. In pools on the south coast the small and deep-rose-red tufts of *Antithamnionella sarniensis*, shown in Plates 38b and 40 (pp. 245, 253), are not un-common.

Larger red weeds in rock pools include the edible dulse, *Rhodymenia palmata* (Pl. 5, p. 64) and also *Delesseria sanguinea*. Both are also frequently attached to the stems of the tangles in the *Laminaria* zone. The former has irregular fan-like fronds (Fig. 26e), often deeply indented, membranous in texture and purple to reddish-brown in colour. *Delesseria* is a beautiful deep-red weed with elongate fronds resembling leaves in their possession of a conspicuous midrib with laterally running veins (Fig. 26f). No red weed is easier to identify but it is largely confined to the deep pools at the seaward margin of the lower shore.

There remain for final mention the intertidal lichens. Above the *Pelvetia* in the splash zone the rock face often bears patches of the black growths of *Verrucaria maura*, while lower encrustations of green are due to *V. mucosa*. The commonest and largest lichen, *Lichina pygmaea*, grows in small tufts up to an inch high on rocks between the top of the shore and mean sea level or a little lower. The individual tufts grow close to one another and may form a continuous covering over many square feet although never over the wide areas occupied by the larger weeds. It withstands the full force of the sea and for that reason is often commonest on the more exposed rocks.

On land the vegetation varies greatly throughout the year, summer foliage giving place to winter barrenness. On the shore there are no such obvious differences ; the rocks remain clothed with their zonal vegetation of wrack at all seasons. The immediate impression is one of perennial vegetation. But closer study reveals constant loss and renewal. The life of individual fucoid weeds probably seldom exceeds two or three years. But they are amazingly resistant to the effects of a wide range of temperature, of exposure and of wave action. They are ripe for long periods and their reproductive cells settle quickly to the bottom and there germinate with the production of little attached sporelings, which are almost always to be found. Life in its early stages appears to be easy, judging by the wide extent of the shore on which these young plants will settle, although, as we have seen, the adults are confined to very well-defined zones. But there is always a copious supply of young plants to repair the inevitable wastage of the older plants which, after the exhausting process of prolonged reproduction, seem to lose strength and are torn apart by the continuous strain of the surging tides. There is also much repair by the older plants with new growth arising from the base to replace torn and broken fronds.

FIG. 27—Holdfast of *Laminaria*. (From Harvey, *The Sea-Side Book*.)

Other weeds, although always present, are much reduced in the autumn and early winter. *Chondrus* and *Gigartina* for instance are at their best in mid-summer when much of the fronds become covered with dark spots where reproductive spores are formed. But in later storms the fronds break off till little but the attachment discs may be left. New growth starts to appear in the early months of the year. Some of this comes from the base of the old plants, some from the sporelings. Some weeds do appear to be strictly annuals because at the end of their particular season they disappear—but not before a liberal period of reproduction which produces sporelings that survive in protected areas or below tide level and form the stock from which the next crop will arise. Examples of these annuals are commonest amongst the many types of filamentous red weeds.

More careful examination therefore does reveal a seasonal change in the plant population of the shore, due partly to the natural dying off of the annuals but also to the destructive action of autumnal and winter storms on plants enfeebled after reproduction. And it is not only the plants of the shore that suffer. The mounds of long stems and torn fronds of *Laminaria* and other sub-littoral tangles that cover the strand line after a heavy storm, representing along outer Hebridean shores many hundreds of tons, are witness to action in the deeper zone they inhabit.

Systematic observations on a shore throughout the seasons will also reveal an upward and downward movement, a migration in fact, of many species of shore algae. Some plants will be found higher on the shore in the summer, others lower. Clearly they react differently to the varying factors of temperature and of light. It is not, of course, the plants that move but their spores, carried freely in the water and able to settle at any height on the shore. Conditions prevailing at the time will control in what regions they can germinate and successfully establish themselves as young plants.

Many of the green weeds, including *Enteromorpha*, *Ulva*, *Cladophora* and *Chaetomorpha*, live in profusion in pools high on the shore in the summer when this region may be green with them. But in the winter they will be found lower down, in the mid-tidal area. Some of the red weeds, such as *Delesseria*, show the same seaward migration in the winter. The most probable reason for this is the fall in temperature. In winter the sea is warmer than the air and so only in the lower pools is the temperature high enough for successful growth by the young

PLATE VII

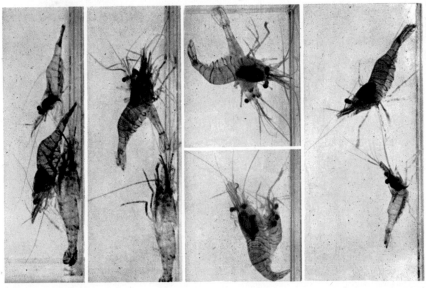

Hans Höglund

b. Photographs under water of pairing in the Common Prawn, *Leander squilla*. In the upper two the cast shell of the recently moulted female is shown (× approx. ½)

D. P. Wilson

a. Skeleton Shrimp, *Caprella aequilibra*, on red weed, *Ceramium* Photographed under water (×5)

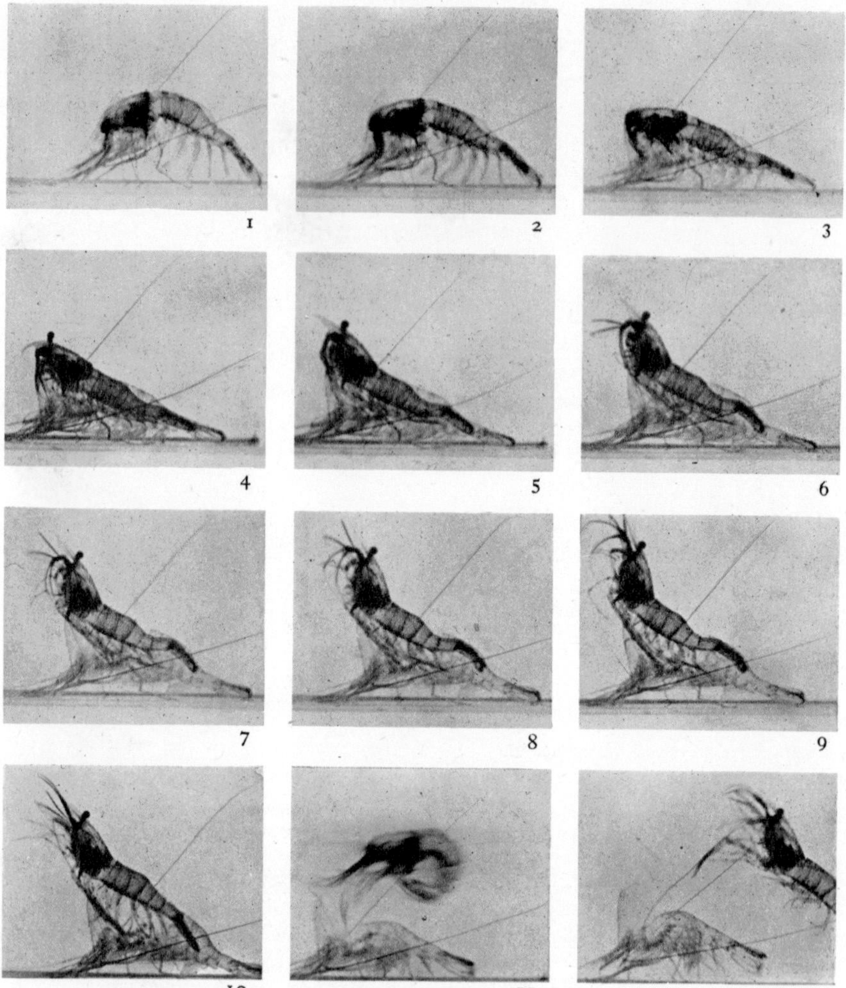

Hans Höglund

Moulting process in *Leander squilla*. The photographs are enlargements from a 16 mm. film strip exposed at a speed of 64 pictures per second. Interval between pictures 1 and 2 and so on to 9 and 10 is 2 seconds, between 10 and 11 it is $\frac{1}{16}$ second and between 11 and 12 only $\frac{1}{32}$ second ($\times$ approx. $\frac{1}{3}$)

plant. Upward movement in the winter is shown by some of the filamentous red weeds, such as *Polysiphonia* and *Ceramium* and also by the brown weed *Scytosiphon*. In these it seems as though the need for greater light " drives " the algae into the higher pools. But it is equally possible that only in the winter months is the water in these pools sufficiently stable in terms of acidity to permit the growth of sporelings of these weeds. The importance of this factor in controlling life within rock pools is discussed in the next chapter.

Finally within a single large pool weeds may move from season to season. Some may live in the shallow, well-illuminated end of a pool in winter but in summer be found only in the deeper and more shaded end. Here light alone appears to be the controlling factor.

Many things remain to be found out about shore plants. Why is it, for instance, that the sporelings so widely cast and so vigorous of early growth finally die off in all but the one zone which may not differ in anything beyond the texture of rock or the angle of slope from adjacent areas where the adult plant is absent ? The reproductive habits of algae are most elaborate ; some alternate between sexual and asexual generations, others, such as the Fucaceae, only reproduce sexually. The rates of growth and of replacement after natural or artificial denudation of the commoner weeds are now being actively studied owing to the increasing economic importance of many of them, but much purely observational work remains to be done by anyone with the interest, care and time to devote to problems the fascination of which grows with added knowledge.

But as a zoologist I have trespassed perhaps too long in the domain of the botanist. The yet more varied field of animal life presents itself for survey now that we know something of the green, brown and red weeds which provide so much of shelter, attachment or food.

CHAPTER 7

LIFE IN ROCK POOLS

"And here were coral-bowers,
And grots of madrepores,
And banks of sponge, as soft and fair to eye
As e'er was mossy bed
Whereon the wood-nymphs lie
With languid limbs in summer's sultry hours.
Here too were living flowers,
Which like a bud compacted,
Their purple cups contracted,
And now, in open blossom spread,
Stretch'd like green anthers many a seeking head."

ROBERT SOUTHEY

THE pools which dot so many rock shores provide a natural introduction to the study of shore life. Here marine animals and plants
may be viewed in their natural medium. An exposed pool is a kind
of natural aquarium with a balanced population of plants and animals ;
it is the microcosm of the sea but with the significant qualification that
whereas the sea is a most constant medium, varying so slightly in
salinity throughout its wide extent and changing in temperature slowly
and within a narrow range, conditions in these pools fluctuate widely
and suddenly. The population, although spared the dangers of
desiccation, must possess those intangible but very real powers of
functional adaptation which have already been stressed as constituting
the fundamental distinction between the inhabitants of the shore and
those of the open sea.

The extremes will naturally be greatest when the pool is high on
the shore and so exposed for long periods ; deep pools near low tide
level differ little from the margin of the sub-littoral zone of which they
are but an extension. But in the higher pools the water is quickly
warmed by the summer sun when the tide is out and then still more
quickly cooled when it returns. In winter, when the air is colder than
the sea, it is the incoming tide that brings warmth. Under a hot sun
the water evaporates and salinity rises. Fresh water, as rain or land
drainage, does not necessarily reduce salinity throughout the pool

because it tends to form a layer over the heavier saline water below and so animals and plants in this deeper layer may be largely unaffected. This layer of fresh water also prevents loss of heat from below. But however the salinity be influenced during the ebb tide, the returning sea restores normal conditions, although in pools above mean high water level of neaps the influence of the sea may be withdrawn for days at a stretch.

An important result of the concentration of a relatively large number of animals and plants in a restricted volume of water is the wide fluctuation in conditions over day and night owing to their chemical activities. During the daytime the plants by means of their green colouring matter, chlorophyll (present in all, but masked by other pigments in the brown and red weeds), are utilising the energy of sunlight to form carbohydrates, i.e. sugars and starches, from carbon dioxide and water. This involves the liberation of oxygen which both plants and animals need for respiration. Within their bodies this oxygen is used to break down organic matter, formed initially by the plants, with release of the energy originally derived from the sun. Such breakdown involves the re-formation of carbon dioxide and water and other waste material containing nitrogen and phosphorus. These are passed back into the water and are available for use by plants, the first two in the formation of carbohydrates, the last, containing the all-important nutrients, for the elaboration of proteins. So the animals and plants together form a closed system, oxygen liberated by the plants being used in respiration by the animals and the waste products from these being utilised in the building-up of organic matter by the plants.

Respiration proceeds continuously, falling off somewhat at night because it is colder then, and all shore animals (except seals and birds) are " cold-blooded," with temperatures practically the same as, and with their activities related to, the surrounding temperature, but the photosynthetic activities of plants are restricted to the hours of daylight. It follows that the amount of carbon dioxide in the water is reduced during the day while oxygen accumulates. In the night the opposite happens: there is a steady demand for oxygen while carbon dioxide accumulates. In the open sea, owing to the enormous volume of water and the much greater dispersal of life, these diurnal changes are minute, but in the confined volume of a densely populated pool they may be very great. On the exposed surfaces of tropical coral reefs there may be seven times as much oxygen in the pools by day as there is during the night. Differences are less in our colder waters but they remain important. Lack of oxygen is not a serious matter except in hot weather because most animals can withstand it for a limited period. It is the variation in carbon dioxide which matters. When this accumulates by night the water becomes increasingly acid and when it is reduced owing to intense photosynthesis by plants in the daytime the water becomes alkaline.

These fluctuations are matters of prime importance to the inhabitants. Animals of the open sea are never exposed to such changes and quickly die if placed in water only slightly more acid or more alkaline than normal sea water which is a little on the alkaline side of neutrality. Life in rock pools involves the capacity to withstand the effects of this wide range in the reaction of the water and so a third functional adaptation is demanded of these shore-living plants and animals.

Such extremes are never permanent on the shore, they are oscillations, though on occasion extreme, about a mean which represents the steadier state of the water of the open sea. Even where this does not come in to redress with a sudden rush the balance that has been temporarily upset, a fundamental balance is achieved : the opposite extremes of night and day cancel out. It was the significance of this relation between plant and animal life which Gosse realised when he established his first balanced aquaria, with adequate but not excessive numbers of both plants and animals, and so started the fashion which was to introduce them into so many Victorian homes.

Pools vary greatly in size, ranging from mere cup-shaped depressions or angular crannies in the rocks (Pl. 6, p. 65) to wide stretches of water which, if near low tide level, harbour a rich growth of the larger seaweeds. They differ also in character, for reasons just considered, according to their height on the shore and the consequent degree of exposure they experience. Indeed they may be said to extend

---

PLATE 7
a. ROCK POOL similar to that shown in Plate 6 but also containing the green weed, *Enteromorpha*, and the Common Periwinkle, *Littorina littorea*

b. CORNISH SUCKER-FISH, *Lepadogaster gouanii*, guarding eggs fixed to the surface of the rock

PLATE 7

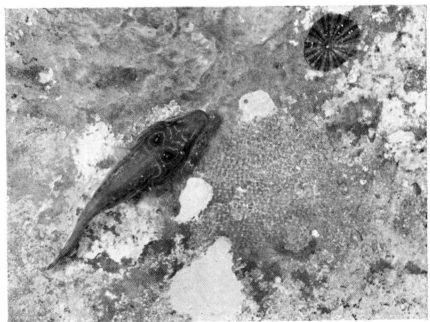

Photographs by D. P. Wilson

PLATE 8

D. P. Wilson

beyond the technical limits of the shore to the splash zone where their contents consist of fresh water mixed with salt spray. Such pools may be encountered first when the shore is approached. Here the seaweeds are exclusively filmy blue-greens and greens, notably *Enteromorpha* (Pl. 2a, p. 5). The animals are confined to such as can withstand the permanent low salinity—because here there is no redress by an oversweeping tide. In other words it is a brackish-water, or estuarine, fauna and is composed largely of small crustaceans with possibly shore-crabs. Such a fauna is better left until we come to deal with the inhabitants of estuaries and with the problems that confront them.

In the truly marine pools little but weeds may at first be apparent. The closer the search, however, the more will be revealed of larger animals lurking in crevices or under stones, many of them toning with the background in virtue of colour or outline, of creatures so trans-parent as to be initially invisible, and finally of increasing hosts of smaller creatures hiding in weeds or growing as encrustations on these or upon the rocky sides of the pool. So the task of description again involves selection, attempting some introduction to all types of animals that may be encountered but concentrating attention on a few, notably sea-anemones, prawns and sea-slugs, which are commonest or best observed in pools.

The walls of the small pools in the bare rock are typically encrusted with pink *Lithophyllum* (Pl. 6, p. 65) with little forests of purplish coralline weed and dark green *Cladophora*. With them are often many snails but these are commoner on the exposed rock and can be left till later. Only under water is it possible to see expanded sea-anemones. The commonest, extending high on the shore and as common under the shelter of exposed fucoid weeds as it is in pools, is the beadlet-anemone, *Actinia equina* var. *mesembryanthemum* (Pl. 8, p. 81). This is usually dark-red with a blue spot, the site of a battery of stinging cells, at the base of each tentacle. There are many colour varieties, some paler or more brilliant red, others of various shades of brown, green or even yellow to orange. There is also the larger " strawberry " anemone, sometimes regarded as a mere variety, *A. equina* var. *fragacea*

PLATE 8
BEADLET - ANEMONES, *Actinia equina*, showing examples of the larger strawberry variety (var. *fragacea*) and several colour varieties of the typical form (var. *mesembryanthemum*)

T.S.S.   G

(also shown in the plate), of red dotted with green, but more probably to be considered a distinct species, *A. fragacea*. The extended tentacles are ready to seize and close over any small animal, even a small shore fish, that comes within range. On the surface of the stout column, the commonest of the little sea-spiders, *Pycnogonum littorale* (Fig. 5, p. 25), with closely applied body and four pairs of outspread legs, may sometimes be found.

The most beautiful of the shore-anemones are the flower-like *Sagartia elegans*, named originally by Dalyell, and *S. troglodytes*, the so-called cave-dwelling anemone. These cannot be adequately described because the colour pattern is so elaborate and also so variable. Gosse described five species on the basis of what we now know to be so many colour varieties of *S. elegans*. Those who are interested must be referred to the beautifully coloured plates in Professor T. A. Stephenson's *The British Sea-Anemones* and to his descriptions. Both species are common and widely distributed around our coasts with *S. elegans* usually near low water level and the other in cracks in the base of pools, as well as in caves, often where there is a thin covering of sand on the surface of which the delicately patterned disk and surrounding tentacles lie outspread.

The largest intertidal species is the dahlia, the scaur-cock of the Yorkshire coast, *Tealia felina* (Pl. 9a, p. 84). This is an altogether stouter animal with a relatively low but very broad column and numerous, rather blunt, tentacles. The colour is somewhat more constant. To indicate the nature of this and also convey some idea of the complexity of pattern found in so many anemones, the following quotation from the description by Professor Stephenson is given. " Area surrounding the mouth grey and pale crimson . . . limited externally by a narrow white ring ; outer part of disk grey, with a strongly marked pattern in rich dark red, made up as follows : The base of each primary tentacle is embraced by two red lines, the space between which, on the oral side of the tentacle, is filled in with red colour on the radius, so forming a red triangle at the foot of the tentacle ; on the aboral side of the tentacle, the red is flanked externally by two opaque cream bands, which run outwards to the edge of the disc. . . . Bases of the secondary tentacles with more elongated triangles. . . . Lip yellowish grey." This elaborately handsome anemone lives typically low on the shore and under stones sheltered from the sunlight. In a pool it may be seen expanded, especially if in the shade, but when touched it contracts under a covering of small pebbles and shell fragments which are held against the column by the wart-like growths which cover it.

An anemone somewhat like *Sagartia* but with a wider disc, more tentacles and, especially in early life, a longer and thinner column is the daisy-anemone, *Cereus pedunculatus*, first described by Thomas Pennant. In general the colour is grey or brown with a variable pattern of lighter coloured spots. In rock pools it lives with the long column insinuated deep into cracks but it is an animal of very variable habits and also lives buried in mud or sand where there are stones for attachment below the surface. The plumose anemone, *Metridium senile* (Pl. 9b, p. 84), is easy to identify because the tentacles are much subdivided, forming a feathery mass encircling the disc. This anemone usually inhabits sub-littoral waters but small individuals occur among weeds, especially *Chondrus* and *Gigartina*, as well as in pools. It may be seen in full glory on pier piles where it forms rich, softly feathered masses of pure white, bright orange or brown. The divided tentacles are unsuited for the capture of the larger animals on which other anemones feed but admirable for that of the minute animals of the plankton which are then conveyed to the mouth by movements of the ciliary hairs on the surface of the disc.

The remaining anemones to be mentioned are largely confined to southwestern shores where the population of these animals is richest, although Scotland boasts of one unique species, found only in Caithness. Of widest distribution is *Anemonia sulcata* (Pl. 10a, p. 85) known, on account of the long tentacles which cannot be fully withdrawn like those of other anemones, as the snake-locks or opelet-anemone. It is dull green or rich brown in colour with mauve tips to the tentacles. It lives in cracks in the bottom of pools but, unlike most anemones, seeks the brightest area and spreads the tentacles towards the light. This habit is certainly not unrelated to the presence in its tissues of vast numbers of microscopic plants, single-celled brown algae known as zooxanthellae. Such intimate association between two organisms (found also in lichens) is known as symbiosis. It is a relationship of mutual advantage. In this case the plant cells gain protection and are able to tap at the source rich supplies of carbon dioxide and especially of nutrient salts needed for the formation of organic compounds. The animals probably profit by this automatic removal of the waste products of their chemical activities. This type of symbiosis is not common in our waters but is widespread in tropical seas where all reef-building corals and the great majority of other coelenterates, together with some other animals, such as the giant clams (*Tridacna*), harbour countless millions of these plants.

Bunodactis verrucosa, the gemmed anemone, is another common species in the south and is usually found in coralline pools. The background colour of the disc and tentacles is green with the column sometimes red but always characteristically dotted with the conspicuous white spots or " gems " responsible for the common name.

The lovely little jewel-anemone, *Corynactis viridis*, occurs around the coasts of Devon and Cornwall. The small and short polyps are usually green although there is the usual range of colour variation, and the tentacles have swollen tips. They usually grow in clusters, as shown in Plate 10b (p. 85), and such a group forms a very beautiful sight. This genus differs in structure from other anemones and there are good reasons for grouping it with the corals.

Two undoubted corals live between tide marks and, although they are also restricted to the south-west and are not common there, they are of sufficient interest to justify description. Corals are normally associated in our minds with the formation of massive reefs in shallow tropical waters, but they also extend into abyssal seas and into temperate and cold waters, although never forming reefs there. As we have already seen, they differ from anemones chiefly in the possession of a limy skeleton and, in most cases, by growth into elaborate colonies. Our two shore species are not colonial ; they are simple cup-corals. Looking down upon their expanded polyps no one would realise that they were not viewing anemones. But on contraction, the walls and radiating internal ridges of the underlying skeleton become apparent. The Devonshire cup-coral, *Caryophyllia smithii* (Pl. 11, p. 92), is common on rocky bottoms at moderate depths off shore and may be found in pools or exposed on a sheltered rock face at extreme low water level of spring tides. The skeleton is about half an inch high and about the same in diameter and is usually white. The animal expands well above this

PLATE 9
a. D A H L I A - A N E M O N E S, *Tealia felina*
b. P L U M O S E   A N E M O N E S, *Metridium senile*, showing some of the many colour varieties

PLATE 9

Photographs by D. P. Wilson

PLATE 10

Photographs by D. P. Wilson

by inflating itself with water and then appears semi-transparent with delicate colouring usually of pink and chestnut.

Our second cup-coral, a beautiful creature of vivid orange or scarlet, has special interest in that it was discovered by Gosse himself in a rock pool near Ilfracombe in 1852. His description of the event, as recorded in his *Devonshire Coast*, is worth quoting. "A very distinct species of Madrepore, and one of great beauty, I discovered to-day. It was spring-tide, and the water receded lower than I have seen it since I have been here. I was searching among the extremely rugged rocks that run out from the Tunnels, forming walls and pinnacles of dangerous abruptness, with deep, almost inaccessible cavities between. Into one of these, at the very verge of the water, I had managed to scramble down ; and found round a corner a sort of oblong basin about ten feet long, in which the water remained, a tide-pool of three feet in the middle. The whole concavity of the interior was so smooth that I could find no resting place for my foot in order to examine it ; though the sides all covered with the pink lichen-like Coralline, and bristling with Laminariae and zoophytes, looked so tempting that I walked round and round, reluctant to leave it. At length I fairly stripped, though it was blowing very cold, and jumped in. I had examined a good many things, of which the only novelty was the pretty narrow fronds of *Flustra chartacea* in some abundance, and was just about to come out, when my eye rested on what I at once saw to be a Madrepore, but of an unusual colour, a most refulgent orange. It was soon detached by means of the hammer, as were several more, which were associated with it. Not suspecting, however, that it was anything more than a variation in colour of a very variable species, I left a good many remaining, for which I was afterwards sorry. All were affixed to the perpendicular side of the pool, above the permanent water-mark ; and there were some of the common *Caryophyllia* associated with them.

" The new species may be at once recognised by its brilliant colours. The whole of the body and disc, exclusive of the tentacles, is of a rich orange, yellower in young specimens, almost approaching to vivid scarlet in adults, especially when contracted, for distension

PLATE 10
a. OPELET OR SNAKE-LOCKS-ANEMONES, *Anemonia sulcata*
b. JEWEL-ANEMONES, *Corynactis viridis*

not only pales the hue, but causes the yellow element to be more apparent. The tentacles, about fifty in number, in my largest specimens, are of a fine gamboge-yellow. . . .

"The form of the calcareous skeleton identifies this interesting addition to the British Corals with the genus *Balanophyllia* of Mr. Searles Wood ; a fossil species of which has been found in the Crag. The royal colours in which the present species is arranged—scarlet and gold—suggest the specific name of *regia*."

Gosse always bestowed popular, as well as scientific, names on his discoveries and the " scarlet and gold star-coral " was not long afterwards found again, this time by his friend Charles Kingsley on the shores of Lundy Island. Since then its known range has been extended around the south coast of Cornwall and on to the coast of Brittany. Although the skeleton is superficially very like that of *Caryophyllia*, it belongs to a very different group in which the skeleton is " perforate," honeycombed with fine strands of living tissue in life, unlike the imperforate skeleton of *Caryophyllia*.

The other common coelenterates of shore pools are the hydroid colonies, but they are not easy to identify from a general description and the more obvious ones can most suitably be left until we come to deal with the fauna of rock and weed. A number of worms will be encountered, including the dark-green *Eulalia viridis*, about three inches long and an active swimmer. It is also common in cracks in rocks, particularly where there are barnacles on which it feeds, and in the spring its eggs, minute green spots within little bags of translucent jelly, are abundant on the shore. On the south coast fortunate discovery may reveal the tube-worm, *Bispira volutacornis*, which is the largest fan-worm of its kind in our waters, and has a magnificent spirally wound crown of tentacles (Pl. 37b, p. 244). It must be observed with caution because even a shadow falling on it causes instantaneous contraction within the parchment-like tube. The ciliated tentacles are the means both of respiration and of feeding. Echinoderms, especially starfishes and sea-urchins, are on the whole commonest under the protective covering of stones near low water mark, and the one most usually found in pools is probably the little brittle-star, *Amphipholis elegans*. It is an inconspicuous grey and not easily seen among the corallines and *Cladophora*.

There is a rich population of crustaceans. Many are minute like the planktonic copepods, others, such as the transparent opossum-shrimps, appear little more substantial than the water they inhabit. One charming little amphipod, the skeleton-shrimp, *Caprella aequilibra*, demands mention ; it is not easy to find but is well displayed in Plate VIIa (p. 76). It lives on corallines and other algae and has a thin body with few appendages, holding on by the hinder of these with the front half of the body free. Like the stick-insect and caterpillars on land, it simulates the foliage which it inhabits.

The prawns call for close attention because they can only be observed in pools and there is much of general application to all of the higher crustaceans that we can learn from them. There are many types of prawns in various levels in the sea but a limited number in pools, the commonest being the large prawns of the genus *Leander* and the small chameleon-prawn, *Hippolyte varians*, with a variety of other forms, less numerous, such as the beautifully coloured *Athanas nitescens* (Pl. 15a, p. 108) which is largely confined to our southern shores. *Leander serratus* (Pl. 12a, p. 93) is our largest prawn and the object of profitable fisheries. Stephen Reynolds in *The Poor Man's House* gives a graphic description of a prawning expedition by inshore fishermen at Sidmouth. *L. squilla* is smaller, about 2 inches long, and is often very common in rock pools.

This prawn spends the winter in the greater warmth and protection of offshore waters but appears in the pools during the spring, often moving into those high on the shore. Then a moderate-sized pool may harbour a hundred or more, but their transparency and habit of hiding among weeds render them difficult to see. But when an arm or stick is moved through the water the backward darting bodies of the prawns quickly become apparent. Once seen they are well worth careful watching. In walking only the three hindermost of the five pairs of legs are employed ; the two first, both armed with fine nippers, are held clear of the bottom ready to pick up fragments of weed or animal debris which are then passed to the six pairs of feeding appendages which manipulate the food into the mouth. Like all of their kind, crabs, hermits and the like, prawns are scavengers, pickers-up of un-considered trifles, and play the indispensable role of such members of any community. Further examination will reveal the means by which they discern their food : the conspicuous eyes mounted on movable stalks and the two pairs of antennae or feelers. Of these the shorter, first pair have each three branches, one extending forward and two upward and slightly backward, while the much longer and undivided second pair are bent back and sweep the ground on either side. This forward, walking motion is gentle ; the animal is undisturbed in the steady search for food. When danger threatens, as from a searching hand or net, there is immediate reaction. The broad tail fan is extended and the whole abdomen bends suddenly under the body, causing the creature to shoot backward with legs raised and with both legs and feelers forwardly directed so that they stream out behind the moving body. At other times prawns may swim. Under the abdomen are five pairs of swimming appendages, the members of each pair connected by little inward projections bearing interlocking spines so that they seem to be holding hands. In swimming these appendages beat in rhythmic series and the prawn moves rather gently through the water with both pairs of feelers and the five pairs of walking legs bent back to give the minimum of resistance to the water. Then, when the animal ceases to swim, these appendages are turned back flat against the under-surface of the tail region, the legs and feelers are extended and the prawn sinks gently down to resume slow perambulation on the bottom of the pool.

Careful searching of a suitable pool with a simple net reveals, with living prawns, numerous cast shells, limp and ghost-like images of their former occupants. About every two weeks in summer the entire covering of the body is cast. This includes the covering of the eyes and of the delicate, feathered gills that lie beneath the overhanging plates on either side of the body, and even that of the stomach, which is armed with teeth for internal mastication and possesses elaborate filters that guard the entrance into a mass of ramifying tubes where absorption occurs. Such moulting is the lot of all crustaceans ; it is the price paid for the protection furnished by an external skeleton which is too stout to stretch as the animal grows. Occasionally, under rocks or in some obscure cranny, a shore-crab may be found emerging from one shell with the new one, appreciably larger, still soft beneath it (Pl. XXIIb, p. 155). At this stage in life even the largest of crabs or fiercest of lobsters is utterly helpless. The body can be squeezed like a sponge and the great claws are soft and useless. Strengthening of the new shell takes but a short time in a prawn but is a matter of some weeks in these larger crustaceans where great amounts of lime have to be incorporated.

The formation of a new shell begins well before the old one is shed and the actual process of casting in a prawn is surprisingly rapid, between 9 and 22 seconds, as shown graphically in Plate VIII (p. 77).

The shell splits transversely across the back at the junction of the jointed abdomen with the shield or carapace that covers the head and thorax. The carapace then hinges upward while still attached in front, most of the appendages are freed and the abdomen is drawn clear of the old covering. Finally the animal makes a sudden upward spring and is free. Water is taken in and the new, soft shell stretches with the surface of the enlarged body. After a short rest the prawn can swim normally, and within a few hours the mouth parts are hard enough for it to feed. At the end of two days the shell is as hard as its predecessor.

In female prawns there comes a time, in early summer, when the ovary is ripe and ready to liberate eggs. Then comes a moult which represents a change to an " egg-carrying " condition. Immediately after this, as shown in Plate VIIb (p. 76), pairing occurs, the female being still soft, but the male, which moults earlier, having a hard shell. Sperm is plastered by him on the under side of the female near the third pair of walking legs where the ducts from the ovaries open. Egg-laying follows quickly ; the female stands motionless resting on the tip of the tail-fan and on the legs. In a steady stream the eggs issue from the two openings and are directed back by the beating of the inner parts of the first pairs of swimming appendages. Gradually they accumulate in a dark mass, shown in Plate IX (p. 96), on the underside of the abdomen and attached to all but the last pair of swimming appendages. A cement, poured out by glands in these appendages, flows round each of the eggs and quickly solidifies to bind them securely to a series of long hairs. These are carried only at this stage and are part of the distinctive breeding dress acquired by the female when she moults into the egg-laying condition. In all, up to 2,500 eggs may be so attached. Spawning takes only a few minutes, subsequent attachment of the eggs up to one hour.

Many of such " berried " females may be found in the summer (on " berried " female crabs the even more numerous eggs are carried in a rounded mass between the undertucked abdomen and the broad under-side of the body). Eventually the young prawns hatch out as small swimming larvae to add to the numbers of the temporary plankton. A few days later the female moults, losing the breeding dress and reverting to the non-breeding condition. Sometimes a second period of egg-laying follows later but almost always with an intervening non-breeding period while the ovary is producing a second batch of eggs. Finally, in the autumn, the prawns move down from the rock pools and seek sub-littoral waters, where they pass the winter.

Scattered over the bodies of these prawns are dark spots, each of them capable of changing in size so that the pattern of the body blends with the background. This property of colour-change is best shown in the chameleon-prawn, *Hippolyte varians*. Apart from its smaller size, usually under an inch, this animal is easily distinguished by the pronounced hump in the middle of the body. But although common it needs to be carefully sought. It bears pigment cells containing three colours, red, yellow and blue, and the colour of the animal changes widely as one of these or a combination of them is spread throughout the ramifying extent of the much-branched cells. If the red pigment alone extends then the animal assumes that colour, or, by a combination of yellow and blue, appears green. In this way the prawn, well justifying its common name, assumes the colour of the particular weed, green, brown, or red, on which it may live. After nightfall the colour changes to a translucent blue which is apparently due to outflow of colour from the pigment cells into the surrounding tissues.

The chameleon-prawn provides the best example on our shores of the property of protective coloration. The pigment cells are controlled by chemical means, by hormones that circulate in the blood-stream and which are produced by endocrine glands resembling the thyroid and pituitary glands in our own bodies, in that they have no duct to the exterior but secrete directly into the blood-stream. These are located in the eye-stalks and are excited into activity by way of the eyes. Thus if a green *Hippolyte* is transferred to a red seaweed the consequent change in the quality of light received by the eye causes, by way of nerves passing from the eye, a change in the activity of the gland. In eventual consequence of this, because colour change is slow and takes up to a week for completion, the green colour disappears and the red pigment spreads through the branches of the pigment cells until the animal blends as effectively with the red weed as it did originally with the green.

With some reluctance we pass from these charming prawns to their larger relatives. The lobster, *Homarus vulgaris* (Pl. 13, p. 100), also moves inshore in the summer and occasionally appears between tide marks. There is always a chance of meeting this handsome blue animal lurking in the depths of a pool near low water level of spring tides. And in the extreme southwest the brown rock-lobster or crawfish,

Palinurus vulgaris, with richly sculptured and spiny shell but without the great claws of the true lobster, may also make rare appearance on the shore. Among the most fascinating of the shore crustaceans are the squat lobsters, of which two species, *Galathea squamifera* and *G. strigosa* (Pl. 14, p. 101), are not uncommon in pools and occasionally under stones low on the shore during the spring and summer. The former is the commoner and is some 3 inches long and usually greenish-brown in colour. The latter, or spinous galathea, grows somewhat larger and is the most handsomely coloured of our crustaceans. The great claws, which it is prepared to use on the slightest provocation, are bordered with rows of sharp spines and the bright-red body, as shown in Plate 14, is dotted and lined with blue. These squat lobsters are essentially crawling animals ; the body is more flattened than that of a true lobster and the legs in consequence more widely separated. Moreover, apart from the great claws, all the four pairs of legs are used for walking, and there are no small pincers on the first two pairs as there are in the lobster. The broad but reduced abdomen only comes into action when the animal is startled, then it straightens out and suddenly bends under again sending its owner darting back through the water. But such movements are always of short duration, unlike those of the lobster, which can swim backward for long distances by alternate flexion and extension of the powerful abdomen.

Crabs are found most commonly under rocks but shore- and edible crabs may be found in pools, where also occasionally spider-crabs, species of *Hyas*, *Inachus* or *Macropodia*, with triangular bodies and far-extending legs (Pl. XXIIa, p. 155), will be encountered. All are usually confined to the deep rock pools near the lowest tidal levels. They are often effectively disguised by growths of weed, hydroids or sponge, sometimes placed there by the crab and held in position by hooked spines. This habit of " masking " is carried a stage further in the sponge-crab, *Dromia vulgaris*. This creature is worthy of mention although it is really a southern species, widespread in the Mediterranean. But it does turn up occasionally on our south-western shores, where the specimen shown in Plate 12b (p. 93) was found. *Dromia* has a broad hairy shell, about 3 inches wide. The last pair of legs are bent upward above the level of the shell and bear nippers with which they hold a piece of sponge over the back. This grows until it comes to cover the shell and fit over it like a cap. Viewed from above, when the crab is motionless with the legs drawn under the body, nothing of the animal is visible—only a mass of sponge, full of spicules, and of all animals the least likely to stimulate the appetite of fish that prey on crabs.

One insect demands mention here, the minute, bluish-black springtail, *Lipura maritima*. Especially in the small pools of the upper shore, well sheltered by surrounding rocks, swarms of these creatures may be seen moving about on the surface film. It is the only shore insect which has this habit, displayed by a diversity of freshwater insects such as the pond-skaters. When the tide returns *Lipura* seeks refuge in cracks in the rocks or it may bury itself in sand.

Snails and limpets have already been mentioned as common in pools but more typical of the exposed regions of the shore. The same is true of the common mussel but not of the larger horse-mussel, *Modiolus modiolus*, or of the rock-pool-crenella, *C. discors*. The former is easily recognised by the fringe of horny fibres round the margin of the very dark shell. Although specimens in rock pools are often larger than common mussels they are only young ones, fully grown individuals may be 6 inches long but they live off-shore, rarely exposed by very low tides. *Crenella discors* is a small mussel, pale-green with radiating lines, and is usually found among weeds.

The loveliest of molluscs are the sea-slugs, not closely related to the land-slugs which they resemble in the absence of a shell. They can be viewed adequately only under water when their feather-like gills and other frilled processes are fully displayed. There are many species, the majority small and difficult to discover owing to irregular outline and resemblance to the background. Broadly speaking they are of three types. The commonest possess a round frill of gills on the back near the hind end, where they surround the anus. The sea-lemon, *Archidoris britannica* (Pl. 15b, p. 108), is the commonest and the largest; it has a tough yellow body mottled with patches of green, purple or red. It is not always easy to distinguish against the background of encrusting sponges on which, almost alone amongst shore animals, it feeds, but it is often more conspicuous on the rocky sides of a pool. All sea-slugs lay masses of gelatinous spawn of a form and colour characteristic of the particular species. That of the sea-lemon, also shown in Plate 15b, consists of frilled white coils. The eggs are

PLATE II
DEVONSHIRE CUP-CORALS, *Caryophyllia smithii*

PLATE II

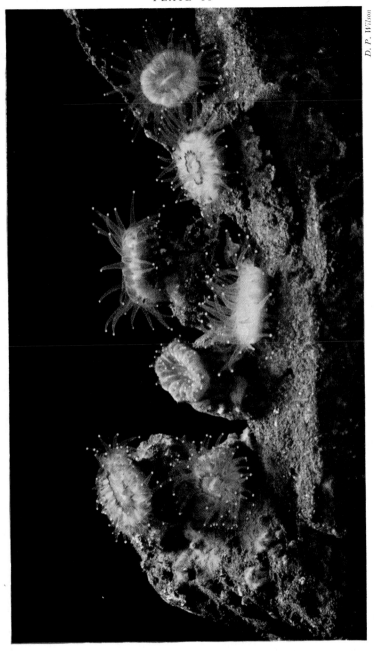

PLATE 12

Photographs by D. P. Wilson

embedded in the jelly and from out of this the larvae eventually emerge to join those of the prawns as temporary members of the plankton. In due course they settle on the bottom, change to the adult form and with the approach of maturity make their way on to the shore for spawning. Life lasts but the one year and ends with spawning whereas prawns live for three or even four years and may spawn five or six times. Every sea-slug lays eggs, it should be noted ; they are hermaphrodite like the land-snails and slugs.

The smaller and more delicate *Goniodoris nodosa* is probably the next commonest species. Here the body is most often white but may be pink or yellow. The texture is more transparent and altogether more delicate than that of the rather coarse sea-lemon and the back is thrown into a series of longitudinal ridges. But it is fruitless to attempt descriptions of the numerous other species, one of which, *Rostanga rufescens*, is shown in Plate 26 (p. 145). Only coloured plates do justice to these animals. The appearance in life cannot be preserved ; colour and delicacy of form and texture are all lost when this is attempted. Fortunately these beautiful animals attracted the attention of Joshua Alder and Albany Hancock, leading members of a notable body of naturalists who made Newcastle a centre of marine biological research in the last century. In the plates of their great *Monograph of the British Nudibranchiate Mollusca*, to which a Supplement was added by Sir Charles Elliot in 1910, the beauty of the sea-slugs is revealed in reproductions of Hancock's paintings. The originals of these remain in the possession of the museum which bears his name at Newcastle.

There remain for mention the two other types of sea-slugs. The first is recognised by the numerous elongated processes borne on the back. The commonest and much the largest is the common grey sea-slug, *Aeolidia papillosa*, the back of which is covered with a soft mat of these thickly set papillae (Pl. 16, p. 109). This is a handsome animal up to three inches long but it yields in beauty of coloration to the much smaller *Facelina curta* of reddish-brown hues and *F. longicornis* with a delicately pink body adorned with papillae of red

---

PLATE 12

a. COMMON PRAWN, *Leander serratus*

b. SPONGE-CRAB, *Dromia vulgaris*, with yellow mass of Sponge, *Ficulina ficus*, held over the top of the carapace

and blue. Also not uncommon but difficult to find because they are
small and live among the branches of hydroids where the egg clusters
are laid, are *Doto coronata* and *D. fragilis*. On the back are carried six
to nine pairs of papillae shaped like fir-cones, those of the former being
pink and of the latter brownish-green, and so large as to appear top-
heavy. All sea-slugs of this type feed on coelenterates, the grey sea-
slug always on anemones, and they accumulate the undischarged
stinging cells of the prey within the cavities of the papillae. The
evolution of such a procedure is baffling but it is usually considered
that, by rendering these papillae unpalatable to fish and other possible
enemies, the stinging cells protect their new owners. In point of fact,
however, all sea-slugs, though slow and naked, appear to be avoided
by other animals.

The last type of sea-slugs are those in which the body is extended
into a series of lateral processes. These are less common on the shore.
The largest one likely to be found is *Dendronotus arborescens*, which usually
only attains full size, some two inches, in sub-littoral waters. The body
bears many branched processes of mixed brown and red and blends
well with a background of seaweed.

Just occasionally summer search in rock pools low on the shore
may reveal an octopus, but the inconspicuously mottled body and habit
of lurking in crevices renders discovery unlikely. These animals are
abundant off shore along rocky coasts. There are two species, *Octopus
vulgaris* (Pl. 17, p. 122), occurring along our southern shore, and
Moschites cirrosa further north. They are very similar but distinguished
by the double row of suckers along the tentacles of *Octopus* and the
single row along those of *Moschites*. When observed in life, either in a
rock pool or in an aquarium, the flow of colour over the surface of the
body is fascinating to observe, the animal paling and then suddenly
flooding with colour. This sudden reaction is in striking contrast to
the slow colour-change in crustaceans, such as the chameleon-prawn,
and the mechanism is correspondingly different. In the octopus the
colour cells consist of little bags containing pigment, and the volume
of these is controlled by the contraction or relaxation of fine strands
of muscle attached to the outer wall. These are under nervous control,
so that the change in size is practically instantaneous.

There is a not unnatural aversion to handling an octopus and
certainly the grip of the suckers on the skin is not pleasant. But these
are harmless: what should be avoided is the mouth which lies in the

depression between the bases of the eight tentacles. This carries a horny beak, not unlike that of a parrot, which can nip through the skin. Like its allies, the ten-armed squids of the surface waters and the cuttlefish of sandy bottoms, the octopus is a predacious carnivore feeding exclusively on living prey, chiefly fish and larger crustaceans.

There remain for mention the shore fishes which are especially common in rock pools. They are worthy of a chapter to themselves when we can also mention other fishes which are especially characteristic of inshore waters and may be found on or very near to the shore.

CHAPTER 8

SHORE FISHES

"I salute you from the land of the mountain and the flood ; from amid scenes worthy the pen of Virgil, and the pencil of Loraine ; from the solitary village of Bethgellert, where the science of Ichthyology recently engaged the attention of your friend ; that interesting science which includes the order, genera, and species of those animals which have either a naked, or scaly body ; are furnished with fins and destitute of feet ; belonging to the fourth division into which Linnaeus has divided the animal kingdom. Their natural history is necessarily involved in more obscurity than that of land animals, from the difficulty of ascertaining their habits, instincts, and specific differences ; yet sufficient is already known to excite the curiosity, and reward the diligence of the naturalist."

MARY ROBERTS : *The Sea-Side Companion.* 1835

T H E typical fish, a cartilaginous shark or dogfish or a bony mackerel or haddock, has a spindle-shaped body admirably fitted for swift movement through water. But fish are one of the supremely successful groups of animals, and different types of them have become adapted for life in the surface and middle waters of the sea and on the floor of the ocean at all depths from the shallow sub-littoral zone to the abyssal depths. Others are equally at home in rivers and lakes and others again, like the eel and the salmon, spend part of their lives in the sea and part in fresh water. Within their respective zones fish have exploited the possibilities of life in every environmental niche and of every available source of food from finely divided plankton to other animals as large or even larger than themselves. They vary as widely in form as in habit. The typical rounded body may become flattened from above downward or from side to side, it may be drawn into attenuated length or compressed into an almost globular mass, all with associated changes in the shape of the head and in the disposition and form of the fins.

A general account of fish lies beyond the scope of this book ; but amongst the environments they have successfully invaded is the shore. A number of fish live in the shallow water around the base of rocks and among the forests of tangle-weeds or beds of eel-grass

PLATE IX

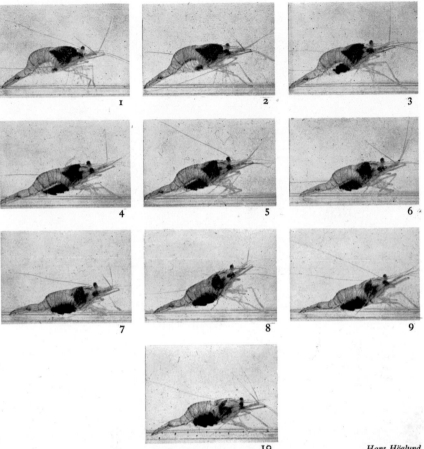

10   *Hans Höglund*

Spawning process in *Leander squilla*. The eggs are seen accumulating under the abdomen
after liberation while the dark mass of the ovary within the thorax diminishes in size
There is an interval of about 30 seconds between each exposure (× approx. ⅔)

PLATE X

D. P. Wilson

a. Male Sea Scorpion, *Cottus bubalis*, beside eggs. Tentacles of the Fan Worm, *Bispira volutacornis*, can also be seen. Photographed under water ($\times \frac{1}{2}$)

D. P. Wilson

b. Rock Goby, *Gobius paganellus*. Photographed under water ($\times \frac{3}{4}$)

and are thus characteristic members of the sub-littoral fauna of inshore waters which extends into the deep rock pools exposed at low water of spring tides. A few other fishes are true members of the intertidal fauna, as fully adapted for shore life as the invertebrate animals and seaweeds which live in the same region. The British fauna does not, unfortunately, contain anything so extremely and delightfully modified as the little mud-skipper, *Periophthalmus*, of tropical shores which can skip about on the exposed surface of mudflats and even across the surface of pools, and is so completely adapted for breathing air that it is drowned if submerged for any length of time ; but exploration of our shores will reveal a number of small fishes only some degrees less adapted to withstand the rigours of life between tide marks.

Here we can do no more than make a brief survey of the commoner of these shore-dwelling and inshore fishes, covering the inhabitants of sandy as well as rocky shores and pointing out such details of structure and habit as are particularly associated with the mode of life. For more detailed description and classification the reader is referred to J. Travis Jenkins's admirably illustrated *The Fishes of the British Isles* and to the classic nineteenth-century works of William Yarrell (*British Fishes*), Jonathan Couch (*Fishes of the British Islands*) and Francis Day (*British Fishes*), while for more general accounts of the structure and habits of fish there are *The Biology of Fishes* by H. M. Kyle and *A History of Fishes* by J. R. Norman.

Pools at and above high water level, brackish rather than truly saline, are often inhabited by the little three-spined stickleback, *Gasterosteus aculeatus*, a typical freshwater and brackish-water species which never extends below the uppermost margins of the shore. The larger fifteen-spined species, *Spinachia vulgaris*, on the other hand, is a marine animal and may be found in intertidal pools on rock and sand or even sheltering under stones and weed when the tide is out. It is about six inches long with a narrow pointed snout and the characteristic fifteen erect spines along the middle line of the back. It is an active and pugnacious animal with particularly interesting breeding habits. During the early summer the male, changing in colour from green to blue, proceeds to the construction of a nest by binding together seaweed with threads formed from its kidneys and then excavating a cavity in this. He next selects a mate, who throughout retains the usual green colour, and she lays a number of relatively large eggs in the nest, which the male guards for some three weeks until they hatch.

T.S.S.   H

This habit of protecting the eggs is found in a variety of other shore-living fishes, although it is unknown among the fishes of open waters, among which the eggs are most usually discharged in vast numbers freely into the sea to float in the surface waters and give rise to larvae that become members of the temporary plankton. But it is a major advantage for the young shore fishes to develop in the region where they will spend adult life and relatively few and large eggs are laid, sheltered from the force of the seas and often guarded, usually by the male parent, from predatory attack.

Both of the common sea-scorpions of rock pools, the father-lasher or bullhead, *Cottus scorpius*, and the long-spined sea-scorpion or cobbler, *C. bubalis*, lay orange-coloured masses of eggs during the winter and spring months (Pl. Xa, p. 97). Each large egg contains abundant yolk so that the young hatch out at a comparatively late stage after an incubation period extending up to six weeks. The spawning habits of *C. bubalis* have been observed by Mr. R. Elmhirst, who writes: "The colours of the male were glowing brilliantly, he jumped and turned about excitedly in front of the female, repeatedly stopped, erected and twitched his fins thus displaying his colours to perfection ; the rapidity of the respiratory act and his general behaviour gave the impression of considerable excitement." Later, after the eggs had been laid, the male ejected semen over them, "jumped and turned about rapidly and excitedly" and several times "darted open-mouthed at the female and engulfed half her head in his mouth."

There is no mistaking these sea-scorpions with their large flattened heads bordered by sharply projecting spines. The body also bears spines but is otherwise smooth and without scales. The father-lasher is the larger and has four large spines on the gill covers whereas the other species has five. Both are typical shore fishes, incapable of swimming for any distance and capturing their food by stealth. The irregular shape and mottled colour render them almost invisible in the pools or among the weed, where they habitually live and from which, by a sudden flick of the large pectoral fins just behind the gill covers, they dart upon some unsuspecting crustacean, prawn or shrimp, which is engulfed in a single snap of the large mouth. When they are handled, these fish still further broaden the head by raising the spiny gill covers and this reaction may save them from being swallowed by sea-birds which can pick them out from under rocks when the tide is out. The flattened belly enables them to cling closely to the surface of rocks and to insinuate themselves within crannies for protection against the force of breaking seas.

The blennies are amongst the commonest and best adapted of shore fishes. Three species occur commonly between tide marks, the shanny, *Blennius pholis*, the gunnel or butter-fish, *Centronotus gunnellus* (Pl. 18b, p. 123), and, in certain regions, the viviparous blenny, *Zoarces viviparus*. All have rather ungainly bodies which taper little towards the tail and bear a continuous or almost continuous fin along the mid-line of the back. The shanny is a yellow or greenish-coloured fish about four inches long and blotched with dark patches so that it blends well with the usual background of irregular rock or weed. Like many shore fishes it has no scales and the body is soft and slimy. It is to be found in shallow rock pools or lurking under stones ; it is said, indeed, actually to leave water voluntarily and bask in the sunshine. The feeding habits are very different from those of the sea-scorpions ; the teeth are broad and used for biting off acorn-barnacles and bivalves from the rocks. This fish spawns in mid-summer when numbers of bright amber-coloured eggs are laid in shelter under rocks or in crevices. The male changes colour to a sooty black except for the prominent whitish lips and in this ferocious aspect stands guard over the eggs until the young hatch.

The hardiness of these blennies and their rather engaging habits which give at least the impression of intelligence render them admirable animals to keep in aquaria where the tompot-blenny, *B. gattorugine*,

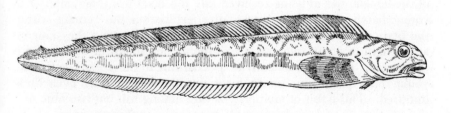

FIG. 28—Viviparous blenny, *Zoarces viviparus*. This fish can attain a length of some two feet. (From Day, *British Fishes*.)

shown in Plate 18c, is also a favourite. This is typically a sub-littoral
species, larger than the shanny and with a pair of fringed tentacles
projecting upward above the eyes. It is a southern species and so is
the little butterfly-blenny, *B. ocellaris*, easily distinguished by the large
fin on the back marked with a conspicuous black eye-spot.

The gunnel is often extremely common under rocks and stones
where it appears to lurk in preference to rock pools when the tide
retreats. As shown in Plate 18b (p. 123) it is an elongate, almost eel-
like, blenny, flattened laterally and so slimy that it slips through the
fingers. It is well named the butter-fish. Some ten to twelve dark
spots are evenly spaced along the back and these, with the elongate
shape, render it most easy to identify. The eggs are laid in small
masses in empty shells, cracks in rocks, or holes made by rock-boring
bivalves, the female collecting them together as they are laid by
looping her body around them. The breeding season extends from
December to March and both parents take their turn in guarding the
eggs, curling their bodies around them so as to prevent their being
washed away by the sea. Incubation lasts for over a month, but
the young larvae then merge into the temporary plankton and spend
several months drifting in the surface waters before some, at least,
make their way to shallow water and so to the shore where adult life
is passed.

The viviparous blenny (Fig. 28, p. 99) is a larger animal, up to
two feet long, and is a northern species which extends along the
northern and eastern shores of Great Britain. Although larger than
either, it is somewhat intermediate in form between the shanny and
the gunnel but the most interesting feature about this fish is its viviparity. The eggs are fertilised, develop and hatch, all within the ovary
of the female, sperm being received into the body from the male. The
young hatch about three weeks after fertilisation but remain within
the body of the female for a further three months before they emerge
at a length of about one and a half inches and identical in form with
the parent. The numbers born vary according to the size of the female,
young ones producing up to forty but the largest ones over three
hundred. This habit of viviparity is rare among fish but its value to a
shore-living animal is considerable because, in striking contrast to the

PLATE 13
COMMON LOBSTER, *Homarus vulgaris*

PLATE 13

PLATE 14

gunnel, the young can immediately colonise the shore. The northern origin of this fish is revealed by the winter breeding season and by its migration into deeper water during the summer months to escape the increased temperature on the shore.

The gobies are among the commonest of shore fishes but need to be carefully sought. They are small and in colour and habit inconspicuous, most of them living on the bottom concealed under stones or in sand. Ten species have been described around our coasts but only the commoner need be mentioned. The most distinctive feature about the short, blunt head is the position on the top of this of the large eyes, so close to one another that in some species they may almost touch. There are two conspicuous fins on the back, while the hinder, or pelvic, paired fins are united to form a fan-shaped sucker on the under-side of the body (Fig. 29, p. 103). The modification of these fins as organs of attachment indicates that speed is of less importance to these shore-dwelling fish than security among the rush of tidal waters. Most gobies are gregarious, a habit which is most obvious in the spotted goby, *Gobius ruthensparri*, the only one which does not live on the bottom but swims in shoals among weed.

The eggs are large and usually pear-shaped. Like those of the blennies, they are laid on the bottom, each one fixed by the narrow end to the under-surface of stones or shells. The breeding habits are interesting. The male finds a suitable place for a " nest " and then proceeds to find a female for whom he fights other males and then guards during egg-laying. She departs immediately this is completed but he remains to protect the eggs and to drive a current of water over them with his fins until they have safely hatched. Both male and female pair with different partners a number of times during the breeding season.

The common or sand-goby, *G. minutus* (Pl. 18a, p. 123), is everywhere abundant in shallow sandy pools and may extend for considerable distances within the freshening waters of estuaries. This fish is two to three inches long and variable in colour and markings. It makes sudden darting movements and then seems to disappear when it comes to rest, so similar is it in colour and marking to the sandy background. The spotted goby mentioned above is about the same

PLATE 14
S QUAT L OBSTE R, or Spinous Galathea. *Galathea strigosa*

size, but of a green or yellowish-brown colour with a conspicuous black spot at the base of the tail. The rock-goby, *G. paganellus* (Pl. Xb, p. 97), which is common in rock pools except in the north, grows to a length of about five inches and varies from yellowish to dark-brown with a pale band along the top of the first fin on the back. It is not gregarious like the other gobies. The black goby, *G. niger*, is about the same size but rather darker and without the pale band on the fin. It is particularly common in estuaries.

Sucker-like attachment by means of the modified pelvic fins is developed to a higher degree in the lumpsucker, the " sea-snails " and the sucker-fish, all of which are highly characteristic members of the shore fauna. The lumpsucker, or sea-hen, *Cyclopterus lumpus* (Pl. XIa, p. 112), is the most remarkable fish likely to be encountered on the shore and quite unmistakable. It grows to a length of two feet but is commonly much smaller and has a rounded, humped body, clearly not adapted for swift movement, and bearing lines of tubercles. On the under-side there is a conspicuous sucker, formed by fusion of the pelvic fins as in the gobies, but much more elaborate and more efficient. Thomas Pennant writes that " By means of this part it adheres with vast force to anything it pleases. As a proof of its tenacity we have known, that on flinging a fish of this species just caught, into a pail of water, it fixed itself so firmly to the bottom, that on taking the fish by the tail, the whole pail by that means was lifted, though it held some gallons, and that without removing the fish from its hold."

The young of this fish are common in rock pools in summer, but the adults only in spring when they breed. The male is the smaller and tinged with red at this period, while the female is yellow or bluish. Vast numbers of eggs are laid, forming a pink layer over stones and weeds. Again it is the males which guard them and often at great personal risk because during the ebb tide they are preyed upon by birds and rats and, during the flood, by other fishes. An even greater danger comes to these fish during the spring storms when even the powerful sucker may fail to hold them and the shore may be covered by lumpsuckers thrown up upon it to die of exposure. The young are as active as the parents are sluggish and very like tadpoles. They swim quickly but hold with the sucker when they come to rest and then wrap the tail round the large head so that they lose all resemblance to a fish.

The so-called sea-snails, *Liparis vulgaris* and *L. montagui*, are northern

fish which live within the arctic circle and are commonest here on
northern shores although extending as far as the English Channel. They
are little yellow fish only a few inches long and differ from the lumpsucker, to which they are closely allied, by the smooth skin and the
longer fin on the back. They breed in the coldest time of the year,
from January to March, when masses of eggs are attached to seaweeds
and to hydroids but usually below low water level and down to a depth
of thirty fathoms. The little suckers or cling-fish, species of *Lepadogaster*,
on the other hand, are members of a group of warm-water fishes, and
the four species found on our coasts extend as far south as the
Mediterranean with Great Britain at the northern end of their range.
They are typical shore fishes, the largest, the Cornish sucker, *L. gouanii*,
shown with its eggs in Plate 7b (p. 80), being about four inches long.
The body is flattened below and the large head prolonged into a

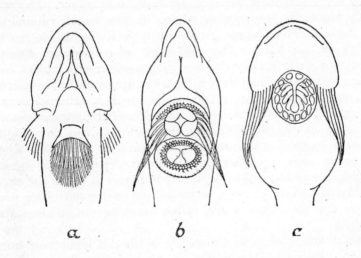

a b c

FIG. 29—Ventral suckers, formed from modified pelvic (hinder paired) fins, in *a.*,
Goby; *b*. Sucker, *Lepadogaster*; *c*. Sea-snail, *Liparis*. Variously magnified. (From
G. A. & C. L. Boulenger, *Animal Life by the Sea-Shore*.)

prominent snout. The sucker (Fig. 29) is relatively larger and much
more elaborately constructed than in the lumpsuckers and sea-snails.
With its aid the fish can grip with a wide surface, while its flattened

form offers little resistance to the rush of tidal waters over the rocks. In consequence of their southern origin, these sucker-fish breed in summer when a single layer of elliptical yellow eggs is attached to the surface of stones or weed and there guarded by one or both of the parents. During the colder months both young and adults migrate into deeper and warmer waters.

The wrasses are common around the margin of rocky shores, among the tangle-weeds and in the deep pools exposed at low water of spring tides. They are somewhat heavily built but handsome and frequently very brightly-coloured fishes with a single large fin running along the back, large scales and conspicuously thick lips. Seven species occur around our coasts, of which it is only necessary to mention the commonest. The largest and usually most frequent is the ballen-wrasse, *Labrus bergylta*, up to two feet long and of bright but variable colours, often red and green. The cuckoo- or striped wrasse, *L. mixtus*, less than half as big, is still more variable and brilliant in colour. The male is usually yellow or orange with five or six blue bands running back from the eye. These bands are absent in the female, which has black blotches on the back. The gold-sinny or rock-wrasse, *Ctenolabrus rupestris*, is a small fish of reddish or golden colour with a black spot at the base of the tail. Then there is the corkwing-wrasse or sea-partridge, *Crenilabrus melops*, another small species which is extremely variable in colour. These fish feed largely on molluscs and crustaceans which they crush up with their powerful teeth and swallow, flesh and shell fragments together. Wrasse have the peculiar habit for fishes of sleeping on their sides and may be observed lying down, especially at night, in aquaria. Most of them build nests in which the eggs are deposited (Pl. XIb, p. 112), although the young may later live for some time in the surface waters before assuming the inshore habit of the adult.

The rocklings, elongated members of the cod family, are not uncommon, especially when young, in pools on the shore. They are easily identified by the projecting feelers or barbels of which the three-bearded rockling, *Motella tricirrata*, has two on the snout and one on the chin, and the five-bearded species, *M. mustela*, two additional ones on the snout. The body is dark and covered with minute scales; careful examination shows that the first part of the fin on the back is reduced to a narrow fringe which lies in a groove. There it maintains a continuous rippling vibration which causes a backward flow of water along the groove, the sides of which carry taste buds which are stimulated by the presence of substances dissolved in the water. Thus the fish, lying quite still on the bottom, is able to detect food in the water in front of it.

The elongation of the body seen already in the butter-fish and in the rocklings is carried still further in the almost snake-like pipe-fishes, which live amongst seaweeds and were characteristic inhabitants of the beds of eel-grass, *Zostera marina*, before disease, to be described later, destroyed these plants. The well-known sea-horse with its deeper body and upright carriage in swimming is allied to the pipe-fishes but is a southern species and only rarely encountered around British coasts.

The pipe-fishes are typical shore and estuarine animals and inhabitants of still waters through which they move slowly by rapid vibrations of the narrow fin in the middle of the back. The head is narrow and elongated with a small mouth through which are sucked the small copepods and other minute planktonic animals on which alone these fishes feed. Each object of food is carefully examined before being swallowed and it is impossible to keep pipe-fishes for long in captivity because they die of starvation if suitable plankton is not available. Their breeding habits are almost unique. In common with the sea-horses and, surprisingly enough, with the small sea-spiders, it is the *male* which carries the eggs from the time they are laid by the female until hatching.

There are three genera of pipe-fishes, *Nerophis*, *Entelurus* and *Syngnathus* ; in the first two the eggs are merely glued in rows along the under side of the body but in the third the male bears a marsupial pouch consisting of two folds of skin that project below and meet to enclose the developing eggs. Even after hatching the young fish remain close to the male parent and may re-enter the pouch for protection. In the sea-horse the pouch, which is tubular, lies at the base of the tail and in it nutrition as well as protection is supplied to the developing young.

Four species of these fishes occur commonly in British waters. Probably the commonest of all is the small worm pipe-fish, *Nerophis lumbriciformis*, about five inches long and with a very short upturned snout. The snake pipe-fish, *Entelurus aequoreus* (Pl. XIIa, p. 113), is a longer animal with a relatively larger snout and often with a rudimentary tail fin absent in the other species. The great pipe-fish, *Syngnathus acus*, is quite an impressive animal if taken at its full length of about eighteen inches ; it has a snout like that of a sea-horse but even longer, while the broad-nosed species, *S. typhle*, is some two-thirds as long and distinguished by the shorter but deeper snout.

The sands have their limited but distinctive fish population. With the little sand-goby already mentioned, there are flatfishes, weavers and sand-eels. But the flatfishes, such as plaice, dabs (Pl. XIIb, p. 113), flounders and other members of the great family of the Pleuronectidae, are only shore fishes for part of their life-history. They spawn offshore in deep water and the eggs and larvae drift in the plankton and are carried into shallow water, where they suffer the remarkable change into the adult form. In the course of this they come to lie on one side, while the eye on what becomes the under-side migrates round the head till it lies beside the other on the upper surface with an accompanying distortion of the bones of the skull. The common flat-fishes mentioned above all lie on the left side, but the turbot and the brill, together with the common topknot, *Zeugopterus punctatus*, which may sometimes be found adhering, sucker-like, to the surface of rocks at low tide, all lie on the right side. These flatfishes blend too well with the background to be easily seen, and are frequently covered with sand except for the upward projecting eyes, but a shrimping net pushed through sand at low water may often bring in a harvest unsuspected by the eye.

The lesser weaver, *Trachinus vipara*, is a true sand dweller. It is a small fish about six inches long, grey in colour above and paler below, with a conspicuous black fin on the back. It burrows rapidly into the sand by means of the pectoral (anterior) paired fins assisted by wriggling movements of the body until only the top of the head, with eyes and mouth clear, and the tips of the spines on the black fin are exposed. It lives largely on shrimps and is common in the sand where they abound. It relies for protection on concealment aided by the poisonous properties of the upward pointing spines. These represent a real hazard to the shrimper who ventures among them with bare feet and this fish has long been notorious. Dr. H. Muir Evans, whose *Sting-Fish and Seafarer* should be consulted by all desirous of a first-hand account of the poisonous glands and properties of this and other fishes, quotes the description of Sir Thomas Browne. In his account of the fish of Norfolk, he writes of " a sting-fish, wiuer, or kind of ophidion or Araneus, slender narrow-headed, about four inches long, with a sharp prickly fin along the back, which often venomously

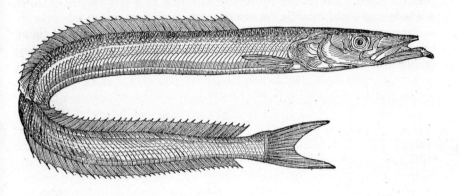

FIG. 30—Greater sand-eel, *Ammodytes lanceolatus.* About twice natural size. (After Yarrell, *History of British Fishes.*)

pricketh the hands of fishermen." There is a second species, *T. draco*, the greater weaver, but this occurs usually in greater depth and is seldom encountered close to the shore. It is about three times as long as the smaller weaver.

The two commonest sand-eels, the greater and the less, *Ammodytes lanceolatus* (Fig. 30), and *A. tobianus* live together in shallow water and may be found burrowing in sand at low water of spring tides. They are respectively eight and four inches long and have smooth eel-like bodies coloured green above and pale below with silvery bands along the sides. In movement they resemble nothing so much as a pointed shaft of silver. The two species are not easy to distinguish but the smaller is the more silvery and tapers more toward the mouth while the fin along the back starts above the pectoral fin in this species and behind it in the greater sand-eel. These fish swim in great shoals and are much preyed upon by other fish as well as by porpoises and sea-birds. Their only safety is in burrowing which is carried out by the aid of the shovel-shaped lower jaw. Owing to the minute size of the scales and the absence of projecting fins—the pectorals are small and the pelvic, or hinder, pair absent—the elongated body slides easily into the sand. Under the protective covering of a few inches of sand a whole vast shoal may lie concealed with no indication on the smooth surface to reveal the wealth of life below. Luck alone will find them but the fortunate digger will be rewarded with a large harvest of highly edible fish.

PLATE 15

a. HOODED SHRIMP, *Athanas nitescens*

b. SEA-LEMON, *Archidoris britannica*, in act of spawning

PLATE 15

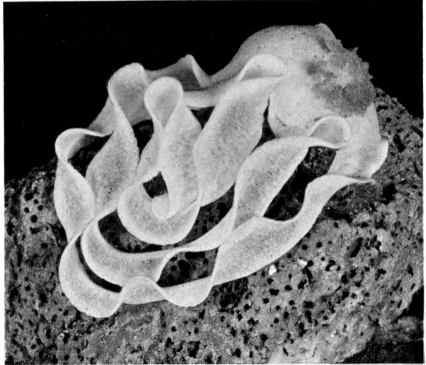

Photographs by D. P. Wilson

PLATE 16

CHAPTER 9

BARNACLES AND MOLLUSCS
OF A ROCKY SHORE

"Patella to the rock adheres,
Nor of the raging tempest fears
The most tremendous power ;
And though assail'd on every side,
Close to her guardian will abide,
Her strength, her fortress, and her pride,
Her never failing tower."

SARAH HOARE : *Poems on Conchology and Botany.* 1831

OUTSIDE the pools there are many habitats for life on a rocky shore. Animals can live on the exposed surfaces of rocks, they can find both protection and, in some cases, food among the dense growths of seaweed, they flourish, well screened from exposure to sun and wind, on the overhanging sides of rocks and within narrow cracks and crevices that run in and between these, or they may burrow within the sand, gravel or mud beneath and around the bases of the boulders.

Conditions are hardest on the exposed rock faces and to colonise these animals must be able to withstand stormy seas and the full rigours of exposure, on the higher shore for hours or even days on end, to the heat of the sun or the effect of drying winds or of frost. It follows that such colonists are the most highly adapted of shore animals. But even they sometimes fail to maintain themselves on rocks fully exposed to Atlantic seas where the violent wash of waters may prevent any effective settlement. In this respect our shore population cannot vie with the encrusting corals and calcareous algae or massively shelled molluscs which flourish in the shattering surf on the seaward crests of coral reefs. It might appear more logical to lead gradually up to this culmination of adaptation to shore life ; on the other hand these exposed animals are the most conspicuous and will be first encountered

---

PLATE 16
GREY SEA-SLUG, *Aeolidia papillosa*, with brown specimens of the Beadlet-anemone, *Actinia equina*, on which, with other anemones, it feeds

by the searcher and for this sufficient reason we will neglect the more logical procedure and start with them.

These fully exposed animals are all encased in stout shells. They comprise molluscs which are characteristically so protected and barnacles which have acquired a shell so similar that, as we have seen, it required tracing of the life-history by Vaughan Thompson to reveal their true crustacean nature. Yet despite their complete adaptation in adult life to fixation on a rocky shore, they retain dependence on the sea which brings them the plankton on which they feed and provides the medium in which the young develop. We can rightly begin with them. They are the commonest of all animals on rocky shores. Their numbers on a well-exposed stretch of shore on the Isle of Man have been estimated as in the neighbourhood of one thousand million over a length of one kilometre. From their water-borne food these barnacles

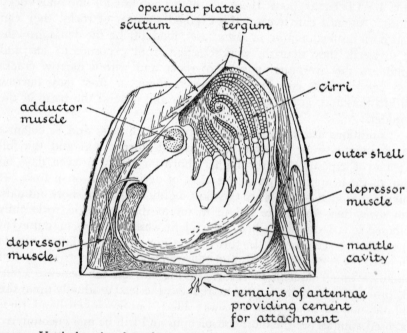

FIG. 31—Vertical section through an acorn-barnacle. The animal is withdrawn into the mantle cavity by contraction of the depressor muscles attached to the opercular plates, the scutal plates of which are drawn together by the adductor muscle. The thoracic feeding appendages (cirri) are folded in as shown. (From *Thomson's Outlines of Zoology*, revised by Ritchie, after Darwin.)

obtain material enough for an annual production of some twelve hundredweight of tissue, weighed after drying. Each year they will liberate some enormous number, up to a million million, of their larval young weighing almost two hundredweight.

There are several species of these sessile or acorn-barnacles. The two that are most abundant, *Balanus balanoides* (Pl. XIIIb, p. 114) and *Chthamalus stellatus*, live highest on the shore and can withstand long periods of exposure. The much larger *Balanus perforatus* extends from the lower regions of the shore into the sub-littoral zone, while *B. crenatus* (Pl. XIIIa) is typically an inhabitant of the latter zone but may occasionally be exposed on the shore at low water of spring tides.

Our prime concern is with the two first of these, both of which may occur in the vast numbers already noted. The first thing is to observe their structure (Fig. 31, p. 110). They are broad-based and firmly cemented to the rock from which a protective wall of six stout limy plates rises, curling inwards above to surround four smaller horizontal plates which form a hinged operculum fitting tightly together and closing the shell when the tide is out. *B. balanoides* and *C. stellatus* are about the same size, both attaining lengths of about one centimetre and it is not too easy to distinguish them. The genera to which they belong are separated by the different arrangement of the outer plates (Fig. 32) but this may be difficult to see, especially in

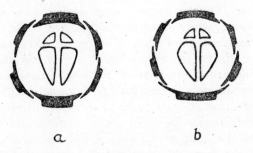

a   b

FIG. 32—Diagrams showing arrangement of shell-plates in the two common genera of acorn-barnacles; *a. Chthamalus*; *b. Balanus*. The opercular plates are shown in surface view. (From Eales, *The Littoral Fauna of Great Britain.*)

older and water-worn individuals. Then the best way of deciding which species we are dealing with is by examining the shape of the opercular aperture which these plates surround. As shown in Fig. 32, this is roughly diamond-shaped in *B. balanoides* but in *Chthamalus* is kite-shaped, with the greatest width not in the middle. These structural details may at first sight appear of but minor significance, and in so far as they concern the economy of the animals this is true, but, as is so frequently the case, they are accompanied by important differences in the habits and distribution of the two species.

When the sea covers them, the opercular plates open to allow the protrusion of six pairs of feathered appendages (Pl. XIIIa, p. 114). These are thrown out and back, in form rather like the partly bent figures of a hand, and make continuous grasping movements, withdrawing into the shell between each. These structures that comb the water, collecting as in a cast-net its minute life, are the walking legs of more typically constituted crustaceans. Barnacles are frequently spoken of as " kicking their food into their mouths." Such change in function is one of the most fascinating aspects of the study of animal life and in this case it represents a fundamental part of the change from freely-moving to sessile existence.

The rate of growth and also the length of life depends on the height that the barnacles live on the shore. On open shores *B. balanoides* occurs roughly between the mean levels of low and high tides. Initially those that settle lower down grow the faster but after the first year the lead in growth-rate is taken by those highest on the shore, which may live for five years or more, whereas the others usually die in the third year. Quicker growth brings earlier spawning to the lower group of barnacles, which ripen in the first year while those higher do not spawn until the second year. The degree of exposure also exerts a considerable effect ; in very sheltered waters few barnacles spawn. Unlike the majority of crustaceans, although in common with many types of sessile animals, barnacles are hermaphrodite. Thus every individual when ripe adds its quota to the vast output of freely swimming young. Fertilisation does not occur in the water but within the cavity between the body of the animal and the protective opercular plates. Here the eggs are fertilised by sperm introduced by the elongated penis of another animal which cannot be more than about an inch and a half away ; hence an isolated barnacle is inevitably sterile. Development takes some four months, at the end of which time the minute but unmistakably crustacean nauplius larvae (Fig. 7, p. 27) hatch out. This occurs in early spring when these larvae are often the commonest animals in the inshore plankton.

PLATE XI

D. P. Wilson

a. Lumpsucker, *Cyclopterus lumpus*, attached by sucker to a stone
Photographed under water (×1)

D. P. Wilson

b. Eggs from the nest of the Ballen Wrasse, *Labrus bergylta* (×3)

PLATE XII

D. P. Wilson

a. Snake Pipe-fishes, *Entelurus aequoreus.* Photographed under water ($\times\frac{1}{2}$)

D. P. Wilson

b. Sand Dab, *Pleuronectes limanda.* Photographed under water ($\times\frac{1}{2}$)

In the water the larvae feed, and grow by a series of moults, the last of which sees the triangularly shaped animal with three pairs of legs assume an oval and laterally compressed form with a bivalved, hinged shell and six pairs of legs. This is the " cypris " stage and the essential intermediate condition between free and fixed life. The cypris does not feed, it is concerned solely with finding suitable attachment. Carried in by the advancing tide it sinks to the bottom and, under laboratory conditions, these larvae have been observed to explore the surface (Fig. 33) for up to an hour, moving perhaps as much as half an inch in this time. A rough surface and shade are the factors which appear to have prime influence on final settlement. For that reason barnacles are most commonly found in grooves, where they are often densely packed, and only relatively sparsely spread over the smooth rock surface. The direction and speed of water movement are both important. If the velocity of the water is too great, barnacle larvae cannot settle. Speeds have not been determined for British species, but in America experiments have shown that the limiting speed for three species varies between half a knot and a little over one knot. But after attachment the animals can withstand a greater continuous speed of water movement. Within the range of movement permitting attachment, the larvae tend to settle with the long axis along the direction of the current. But later they often twist round through an angle of 90° so that the feeding net is cast across the direction of the current. This permits continuous feeding because the currents are usually tidal and so flow first in one and then in the opposite direction and if the barnacles retained their original position feeding would only be possible when the current flows into the concavity of the extended appendages.

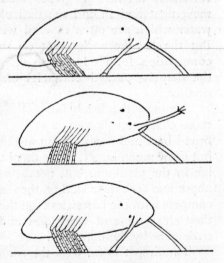

FIG. 33—Successive movements of the cypris larva of an acorn-barnacle when " selecting " a site for attachment. Greatly magnified. (From Visscher, *Bulletin U.S. Bureau Fisheries*, Vol. XLIII, p. 246.)

But we have run ahead of our story. The actual process of attachment is brought about by means of a cement produced by glands within what were, in the freely-swimming nauplius, the sensory feelers in front of the head. After this a complicated series of changes in the arrangement of the different parts of the body leads to the assumption of the adult form firmly cemented to the rock and protected around and above by limy plates.

In general, barnacles flourish best under exposed conditions and can range further up and down the shore owing to a greater tolerance both of exposure and of immersion—because the latter is just as unfavourable as exposure to the truly intertidal *B. balanoides* and *Chthamalus*. Where the two compete, and we shall see later when discussing the distribution of British shore animals that they do so only in certain areas, *Chthamalus* extends the higher—in very exposed regions even into the splash zone, although it remains difficult to understand how sufficient food can be obtained from the sea under such conditions. The probable causes of this " preference " for exposed conditions are first the greater food supply where there is much water movement and second the difficulty of settlement on rocks in still water which are often covered with a thin layer of mud or slime. But the few barnacles that do settle in sheltered areas have little competition for food and so often grow abnormally large whereas the densely packed barnacles on more exposed rocks are usually smaller.

Amongst the intertidal molluscs of rocky shores limpets are the most characteristic and most highly adapted. They have the same broad base and conical form as the barnacles among which they live, but grow much larger. Like them they settle on the shore after a short life in the plankton, but, not being attached, they are able to move about and seek their food on the shore. In no sense, therefore, do they compete with the barnacles ; on the contrary by browsing on the rocks they clear these of vegetation and render a wider surface open to settlement by the barnacles.

The limpet form, with the conical shell and broad foot, has been adopted independently many times in the evolutionary history of the snails. There are several different types of limpets on our own shores and yet another, allied to the land-snails, in fresh water, while in foreign seas there are a variety of others. The common species on British shores belong to the genus *Patella*, in which the stout opaque shell is

PLATE XIII.

D. P. Wilson
a. Acorn Barnacle, *Balanus crenatus*, with opercular plates withdrawn and feeding limbs extruded. Photographed under water (×5)

D. P. Wilson
b. Common Acorn Barnacles, *Balanus balanoides*, showing the typical appearance of rocks in the barnacle zone when exposed (×⅔)

PLATE XIV

a. Common Limpets, *Patella vulgata*

D. P. Wilson

D. P. Wilson
b. Small Periwinkles, *Littorina neritoides*, in cracks in the rock surface in the splash zone
(×1)

ribbed from apex to margin. The three species are not easily distinguished but it can safely be said that most found exposed in the upper part of the shore, although they also occur in pools, are common limpets, *Patella vulgata* (Pl. XIVa). The other two species, *P. depressa* and *P. intermedia*, the latter being confined to southern and south-western shores, live lower down although the latter may occur in pools higher up. The differences in the shell are not very helpful since this varies in *P. vulgata* itself, as shown below. There are also differences in the colour of the foot and mantle, which are darker in *P. vulgata*, and in that of the tentacles which are greyish in this species and white in the other two. The best distinction resides in the character of the teeth on the lingual ribbon but this has to be removed and examined under the microscope.[1]

It is thus best to confine attention to the unmistakable common limpet exposed high on the shore. Uninteresting as it may at first sight appear, no animal is more beautifully adapted in so many ways and it will amply repay careful study. After a cautious approach a sudden kick will usually dislodge an animal but if first tapped gently a succeeding kick, however hard, usually has no such effect and the stout shell will frequently break before the limpet releases its hold on the rock. This is exactly what happens in a stormy sea. The harder the seas pound upon the rock the firmer does the limpet cling. The broad-based conical shell permits the broadest of attachment with the minimum of resistance to the water.

Attachment is by means of the flat rounded foot which is equivalent to the elongate surface on which a land-snail makes slow progress. The mechanism of attachment remains somewhat of a mystery. Initial examination would indicate that it acts as a sucker, creating a vacuum in its centre so that it is held in place by the pressure of the atmosphere. Experiment reveals, however, that the force needed to dislodge a limpet is greater than this while animals may firmly grip small pieces of rock where the foot could hardly act as a sucker. Some other, or additional, means of adhesion is indicated. But the result is clear enough—the capacity to remain attached although fully exposed to the stormiest seas.

Next there is the problem of desiccation. Here again the ability to pull the margins of the shell firmly down against the surface of the

[1] For full details about these species of *Patella* the interested reader is referred to " Studies on the Biology of British Limpets " by R. G. Evans (1947), *Proc. Zool. Soc. Lond.* 117 : 411-23.

rock is of prime importance, but this would nevertheless be useless if water were not retained in the narrow groove which examination will reveal between the central foot and the margin of the shell. Into this groove project the numerous thin leaflets of the gills (Fig. 34). The

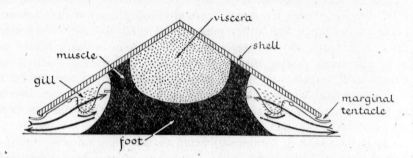

FIG. 34—Diagrammatic section through a common limpet, *Patella vulgata*, showing principal structures and the direction of the respiratory current created by the gills which lie within the groove between the foot and the margin of the mantle and shell. (After Yonge.)

limpets have lost the typical plume-like gills possessed by most types of marine snails and acquired a new type of gill more suited to their peculiar mode of life. If the animal is to respire while the tide is out, indeed if the delicate gills are not to be damaged by desiccation, water must be retained around them.

The rock surface is seldom perfectly flat and yet the margin of the shell must make perfect contact with it if water is to be retained when the tide falls. To accomplish this the limpet has either to grind the surface of the rock to fit the margin of the shell or else make the margin fit the irregularities of the rock. It may do either of these things. If an exposed limpet be dislodged from a soft rock it will be found to have occupied a ring-shaped depression on the surface of this. Similar rings, formed by previous generations of limpets, will be found dotting the rock surface. The margin of the ring is deeper than the centre because the rim of the hard shell cuts deeper into the rock than the more gentle, but frequently quite appreciable, action of the foot. On hard rocks the result of this interaction between shell and rock is to wear away the former so that it comes to conform with the configuration of the rock.

In one way or the other, therefore, the limpet achieves perfect contact between the shell and the rock on which it lives—but only at one place which represents its " home." The animals browse on the algal vegetation and must therefore be able to regain this home whenever the need for avoiding desiccation arises. So the behaviour of the animal comes into this complicated picture of adaptation. Limpets possess an undoubted " homing instinct." The exact nature of this has so far defied analysis ; it does not seem to reside within the restricted powers of sight, smell or touch. As shown in Fig. 35, limpets browse in a rough circle around their homes, travelling at most three feet from this and usually very much less, and are able by this sense of direction to regain their home when it is necessary. Having done so they shuffle round, when probably most of the grinding of soft rock by the foot is produced, until the shell margins just fit the rock below. Finally the powerful muscles which run from the shell down into the substance of the foot, shown in Fig. 34, are contracted and the shell makes perfect contact with the rock.

In general it can be said that limpets move about when the tide is in and it is not very rough, but they also do so to some extent when

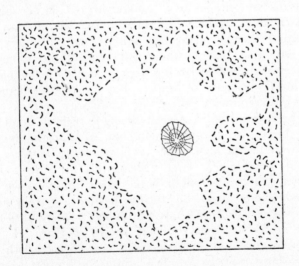

FIG. 35—Browsing area of a limpet, *Patella vulgata*, surrounding algal-covered regions stippled. (From Moore, *Proceedings Malacological Society*, Vol. XXIII, p. 116.)

exposed if the rock is damp and the sun not too hot. On dry rocks each will be firmly secured in its appointed home. When the limpet is browsing, the small head with its pair of tentacles is exposed beyond the margin of the shell and the body swings slowly from side to side. This permits a broad scraping by the lingual ribbon or radula, the characteristic molluscan feeding organ, which consists of a long strap bearing rows of microscopic horny teeth.

The bulk of this lies coiled in the body but the part in use can be protruded from the mouth on the surface of a muscular pad over which it is pulled to and fro like a rope over a pulley. In this way encrusting vegetation of small algae and lichens is rasped off and swallowed. As the teeth are worn away, and the fine scratches which can sometimes be seen on the rock after a limpet has passed indicate how hard the wear must be, new rows come into service because the ribbon is being continuously added to behind so as to make good wear and tear in front. Meanwhile a gentle current of water is drawn in by the ciliated gills beneath the raised margin of the shell. Waste matter collects in the middle of the groove on the right side from whence it is from time to time ejected following a slight muscular contraction.

Limpets are very numerous in suitable areas (182 per square metre have been collected at Port St. Mary in the Isle of Man) and the effect of their browsing on the vegetation is naturally very great. It has been estimated that in their first year of life something like 75 square centimetres of encrusting weed is needed for the maintenance of every cubic centimetre of limpet. But the clearest demonstration is the effect of removing limpets from an area of rock which has previously been almost bare of weeds owing to their constant browsing. Then sporelings can establish themselves so that the rock quickly becomes covered with such weeds as *Enteromorpha*, *Porphyra* and *Ulva* while fucoids may appear later. Meanwhile the surrounding areas where limpets still browse remain bare. These limpets slowly encroach on the weed-covered patch but fucoids beyond a certain size cannot be eaten down and become firmly established. The results of a similar experiment in the *Gigartina* zone are shown in Plate VI (p. 69) in which a boulder cleared of limpets and now covered with the weed is shown in sharp contrast to others where the limpets were left and the summits of which are bare, every young weed being scraped off by the animals. Were it not for the presence of limpets, the rocks

PLATE XV

a. Rough Periwinkles, *Littorina rudis* ($\times \frac{1}{2}$)

b. Common Periwinkles, *Littorina littorea* ($\times \frac{1}{2}$)

PLATE XVI

M. A. Wilson

a. Top-shells, *Osilinus lineatus* ($\times \frac{2}{3}$)

D. P. Wilson

b. Dog-whelks, *Nucella lapillus*, with egg capsules attached to rock ($\times \frac{1}{2}$)

would carry a much greater mass of weed ; on the other hand were the weed not abundant in other regions where it is too large for the limpets to eat away there would be no constantly replenished supply of young plants for the nourishment of the animals.

Unlike the sea-slugs and the land-snails and slugs, the majority of marine snails are of separate sexes. But there is clear evidence that young limpets are mainly males and the older ones mainly females ; in other words that many must start life as males and then change to females. Such a state of affairs is by no means uncommon among marine molluscs, especially in the bivalves where alternation of sex may also occur. The native oyster, for instance, usually starts life as a male, then changes to female and after that alternately produces sperm and eggs at successive spawnings. Limpets breed over the winter, chiefly in January and February, and the sexual products are liberated freely into the sea, a primitive state of affairs not found in the majority of shore-snails. Development leads to the forma-tion of minute freely swimming larvae which live as temporary plankton before settling on the shore. Growth is then steady, a length of about one inch being reached at the end of the first year, animals one and a half to two inches long being two years, and those over two inches, three years old.

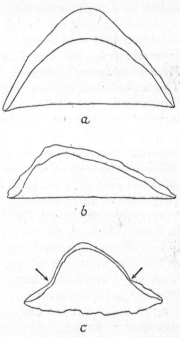

There are notable differences between the shape of the shell in animals which have grown on the exposed rock and those that inhabit pools. In the former the shell is usually high with a relatively narrow base and in the latter it is low and broad (cf. Figs. 36a and b). This

FIG. 36—Shell shapes, in vertical section, of limpets, *Patella vulgata* ; *a.* from high water level ; *b.* from low water level ; *c.* " ledging " of shell after transfer from exposed rock surface to a pool, arrows denoting limits of shell when transferred. About natural size. (From Moore, *Proceedings Malacological Society,* Vol. XXI, p. 213.)

has been related by Professor J. H. Orton, to whom much of our knowledge of limpets is due, to the different habits of the animals. Those exposed have frequently to pull the shell hard down on to the rock to resist wave action or to escape drying up. This continual downward pull by muscles which are attached some distance from the margin of the shell (Fig. 34, p. 116) affects the form of the shell by pulling inward the margins of the mantle tissues which add to the shell in growth. Animals in pools do not need to contract these muscles to the same extent and so there is no such constraint on the lateral extension of the shell. When animals which have grown exposed with sharply conical shells are transferred to pools the effect of this on subsequent growth is shown by a pronounced ledge all round the shell (Fig. 36c, p. 119).

It will be most convenient at this stage to mention the other types of limpets that may be found. In the higher rock pools, never on the exposed rock, the broad and smooth-shelled tortoise-shell-limpet, *Patelloidea (Acmaea) testudinalis*, is not uncommon especially in the north. This is about half an inch long and easily distinguished by its patterned brown shell. It is a less specialised animal than *Patella* in that it retains the single plumed gill. A closely related species with a white and pink shell, *P. virginea*, is more widely distributed and occurs lower on the shore under stones or in the *Laminaria* zone. Both are few in numbers and inconspicuous in size compared with the species of *Patella* but species of *Patelloidea* are large and immensely abundant in the Pacific. They flourish there probably owing to lack of competition from the more highly specialised *Patella* which does not occur in that ocean. In South Africa, on the other hand, where colder Atlantic and warmer Indian Ocean waters bathe the western and southern coasts respectively, no less than 11 species of *Patella* occupy zones on the shore.

A very lovely little limpet, more closely allied to *Patella*, occurs attached to the holdfasts and stems of *Laminaria*, or *Saccorhiza*, usually lying in hollows it has eaten out of them. This is the blue-rayed, or blue-spotted, limpet, *Patina pellucida* (Pl. 19a, p. 126), with a translucent shell dotted with radiating rows of blue spots which appear green when light strikes them at certain angles. After leaving the plankton the young may settle on the fronds of *Laminaria* and develop an elongated low shell or else within the shelter of the holdfasts where the shell becomes rough, round and usually high. Those on the fronds migrate downward in late autumn and so, when they live on *L. digitata*, move away from the terminal regions which break away about this period after fruiting. On rocks, also low on the shore, occurs the small key-hole-limpet, *Diodora (Fissurella) apertura*. It is not easily distinguished against the rough rock surface, but there is no danger of confusing it with the other limpets because of the key-hole opening at the apex of the shell. It is the most primitive snail in British waters, because it retains the two feathered gills which we have good reason for thinking were possessed by the original molluscs from which the modern snails evolved. Water is drawn under the shell at each side of the head and passes out through the opening above.

It is most suitable at this point to make brief mention of the small chitons or coat-of-mail-shells. The commonest, *Lepidochitona cinereus*, occurs on and under stones ; it is usually not more than about half an inch long and far from conspicuous. It is unfortunate that our shores are not inhabited by more massive species such as those, ranging from three to eight inches in length, which abound on tropical shores and around temperate coasts in the Pacific. No molluscs are better adapted for life on the irregular surface of rocky shores. The eight overlapping shell plates (Fig. 14, p. 33) are articulated so that the oval body can be bent readily and the long foot cling firmly by con-forming to the outlines of the most uneven surface. The plates are embedded in a tough tissue which extends around them so that perfect contact with the rock can be maintained when the tide is out. Gills lie in the narrow groove that encircles the foot, but, unlike the small gills of the common limpet, these are feathered structures and represent a multiplication of the primitive pair. The small head cannot be seen until the chiton has been pulled off the rock and studied from below when it appears as a small rounded area, with the mouth opening in the middle, lying in front of the foot. It bears neither tentacles nor eyes. These animals move very slowly and feed, as do limpets, by rasping the encrusting vegetation with the long radula. They are reported to move about at night but to return to their original position early in the day, being possibly guided there by the mucous secretion left on the outgoing trail. When removed from the rock, the body curls round, in much the same manner as that of the black wood-louse or pill-bug. This reaction is probably of service to the animal should it be knocked off the rocks and rolled about in a stormy sea.

We now pass from animals with the limpet form or habit to characteristic snails with coiled shells. Three kinds of these are widely distributed on all rocky shores, namely the periwinkles, the top-shells and the dog-whelk. Four species of periwinkles inhabit our shores. Especially in exposed areas where, in calm weather, it may not even be wetted by spray for days or sometimes weeks on end, lives the small periwinkle, *Littorina neritoides* (Pl. XIVb, p. 115). Occasional individuals may be found down to the middle of the shore but it is a characteristic inhabitant of the splash zone and usually most abundant on slopes facing south. In the crevices of the rocks it finds protection from the force of the sea and is of all shore animals the most highly adapted to withstand desiccation. Like all periwinkles, it is herbivorous and feeds largely on lichens. Despite contrary statements in many books, it liberates spawn into the sea and does so over the winter, from September to April. Spawning apparently occurs rhythmically every fortnight and so coincides with the periods of spring tides which are the only times when the animals are submerged especially during winter months when the height of the tides may be increased by storms. The larvae spend a long time in the plankton before settling in the relative exposure of the barnacle zone and changing into the adult form. They then move up beyond the limits of the shore where adult life is spent.

The rough periwinkle, *L. rudis* (Pl. XVa, p. 118), also inhabits the higher zones, from about half-tide level up to the splash zone, though never as high as *L. neritoides* or in such exposed areas. It is a larger species with a ribbed shell of very variable colour which may be white, yellow, brown or black. Its most interesting adaptation is viviparity although recently specimens with somewhat different shell characters have been reported laying eggs, indicating a possibly distinct species. Periwinkles, like most marine snails with the exception of the more primitive limpets and top-shells, do not discharge their sexual products for chance fertilisation in the sea. The males possess a penis and impregnate the females so that fertilisation is internal. To do this the male mounts the shell of the female until it finds the correct position on the right side of the aperture. Males often mount shells of other species of periwinkles and also of other males of their own species, so that actual copulation and consequent fertilisation is the result of trial and error. Fertilisation within the body of the female has the great advantage that, as in the small periwinkle, it allows for the formation

PLATE 17
OCTOPUS, *Octopus vulgaris*

PLATE 17

PLATE 18

Photographs by D. P. Wilson

of a protective case around the spawn before this is liberated. In *L. rudis* it further permits complete development of the fertilised eggs within the body of the female, from which there eventually emerge numbers of small, shelled young resembling the parent in all save size. The characteristically rounded opening of the shell in this species is possibly connected with easy outward passage of the young. These proceed at once to colonise the shore. Thus *L. rudis* depends on the sea much less than *L. neritoides* ; though the latter often lives on practically dry land. The young of *L. rudis* do not have to undergo the hazards of planktonic life or face the subsequent chances and difficulties of gaining the splash zone should they survive these. On the other hand the possibility of wide distribution by water movements has been sacrificed by *L. rudis*.

The third species, *L. littoralis* (*obtusata*) (Pl. 20, p. 127), is confined to the zone of the large fucoid seaweeds, notably *Fucus vesiculosus* and *Ascophyllum*. The rounded shell of this flat periwinkle may be very like the bladders on these weeds. Protective value has been attributed to this resemblance, though *L. littoralis* generally has an even wider range of colour than the other periwinkles ; most commonly bright yellow, it may also be olive-green, brown, black or striped. It depends on the weeds for food and for the damp atmosphere with which, unlike the two preceding species, it must be surrounded when exposed. Eggs are laid in gelatinous masses on the surface of the weed, and from these crawl out in their appointed time the shelled young. Such a state of affairs represents a compromise between conditions in the other two species. The developing eggs are exposed to greater danger from desiccation, from wave action and from other animals than are the fully protected young of *L. rudis*. On the other hand, the young periwinkles that do emerge safely have not to undergo the difficulties and dangers experienced by the larvae of *L. neritoides* while in the plankton or later by the newly settled young when searching for their final home.

The common periwinkle, *L. littorea* (Pl. XVb, p. 118), is the largest,

---

PLATE 18

a. S A N D - G O B Y, *Gobius minutus*

b. B U T T E R - F I S H   O R   G U N N E L, *Centronotus gunnellus*

c. T O M P O T   B L E N N Y, *Blennius gattorugine*

commonest and most widespread of the four. The shell is usually black but may be brown or red ; young ones are ridged but fully grown shells are usually almost smooth. Within its vertical range, which may be from about high water of neap tides to mean low water of spring tides (although it is not infrequently more restricted), no shore animal has succeeded in solving the problems of existence under such widely differing conditions. It lives on bare rocks or among weeds and stones, also on gravel that may grade into soft mud, and even on sand. It lives equally well when fully exposed to the open sea or protected in sheltered bays ; it extends up estuaries indifferent alike to the lowered salinity and the frequent presence of pollution. Although collected in enormous numbers for food the common periwinkle is so successful a species that it remains everywhere abundant.

L. littorea resembles the small periwinkle in its complete dependence on the sea during early life. Spawning takes place in the south of England mainly from February until about the middle of April. It occurs chiefly at night and on the flood tide when the females liberate numbers of egg-capsules each containing three eggs. From these emerge the small, motile larvae which, after a short life in the plankton, settle on the shore as young periwinkles during May and June. Growth is rapid ; when eighteen months old they are some two-thirds of an inch high and sexually mature although many do not spawn freely until the third year of life. Eventually they may attain to heights of an inch and a half, but maximum length of life is difficult to determine. Growth is slowed down during the spawning period but some periwinkles are parasitised by worms which destroy the reproductive organs and prevent spawning and in these growth is unchecked.

The periwinkles do not rely, like the limpets and chitons, on close adhesion to protect them from wave action. They seek safety by going with the current, not by resisting it. The stout shell wound tightly round the central pillar or columella can be rolled about by the stormiest seas without damage. When this happens the animals withdraw first the head and then the foot into the cavity of the shell and finally close this by a rounded, horny operculum which is carried on the upper surface of the hinder part of the foot. This turns in, as on a hinge, when it is withdrawn so that the round door fits tightly a little distance within the outer margins of the shell. When exposed by the tide periwinkles of all types usually seek shelter in crevices or under weed, often remaining attached by the foot. But the common peri-winkle may frequently be found attached to the sides of boulders, especially on hot days. If pulled away the animal will be found withdrawn into the shell and secured by a thin film of mucus which covers the mouth of the shell. This is produced, probably as a direct response to the dry atmosphere, while the animal is still attached by the foot from which the mucus is produced. The wet mucus spreads over the surface of the rocks where it hardens and so glues the margins of the shell to them. The foot is then withdrawn, the operculum fits into place and the animal remains secured in this manner. Great numbers of periwinkles may be found so attached and all, it will be observed, are arranged in the same way with the lip of the shell uppermost and the spire of the shell below. In this position, with the major weight of the animal resting against the rock below, the film of mucus can hold the shell in position. If, as occasionally happens, a periwinkle attaches itself with the spire of the shell uppermost, it usually topples over when the foot is withdrawn. Even normal attach-ment is very delicate and a light touch, or a sudden rush of wind, will knock the periwinkle off.

The periwinkles are true inhabitants of the shore. They are probably the best modern examples of animals which are moving from the sea on to the land. While only the common periwinkle occurs below low water spring tides, and that but rarely, the small periwinkle when adult is really a terrestrial animal, while the rough periwinkle is almost equally exposed. In agreement with the change in habit, the gills become reduced. The two species lowest on the shore, *L. littorea* and *L. littoralis*, both have the normal gill within the cavity behind the head through which water circulates. But in *L. rudis*, and still more in *L. neritoides*, the gill is reduced while the cavity becomes more vascular and both in structure and function more like the " lung " into which the gill chamber is converted in the true land snails. Both species are capable of obtaining oxygen from the air ; this is essential to animals which may be submerged only during the fortnightly spring tides or even less frequently.

The top-shells are as characteristic inhabitants of rocky shores as are the limpets, to which they are more closely related than the periwinkles. Unlike the latter none of them extend on to soft substrata of mud or sand ; this has to do with the primitive nature of the gills which easily become clogged and useless if much suspended matter is carried into the gill chamber. But they are to be found on all rocky shores including Indo-Pacific coral reefs where the large *Trochus niloticus*, three to four inches in diameter, is collected for the sake of the shell which is cut into pearl buttons. Like limpets, the top-shells rasp the vegetation from the rocks and the eggs are fertilised outside the body, but they resemble the periwinkles in the possession of a stout coiled shell which, with the animal withdrawn and the opening closed by the horny operculum, can be rolled about without suffering damage.

On our shores are a number of species inhabiting successive zones although seldom in such sharply defined regions as the different species of periwinkles. They also extend into the sub-littoral zone where additional species occur which are seldom found between tide marks. One species, largely confined to the south and west coasts, does however often occur in a more limited zone than any other shore animal. This is the large *Osilinus lineatus* (*Monodonta crassa*) (Pl. XVIa, p. 119) of mottled grey colour but with the rounded apex usually worn to expose the silvery layer below. It is sometimes confined to a narrow zone, only a few feet wide, between the levels of extreme and mean high water of neap tides, although it may extend over a wider area as shown in Fig. 60 (p. 197). Discussion of the many interesting problems involved in zoning is delayed until later when the reader has a better knowledge of the plants and animals involved. The flat top-shell, *Gibbula umbilicalis*, is also commonest in the south. It ranges from the lowest tidal levels up to the top of the region occupied by *Osilinus*. It is shown with the grey top-shell or " silver tommy," *G. cineraria*, in Plate 21 (p. 128). This species is more widely distributed but its range on the shore is more restricted, it never extends so far up this as the flat top-shell and descends into the sub-littoral zone. These two species are much alike with flattened, grey shells with oblique markings, somewhat broader in *G. umbilicalis* which also has a deeper depression, or umbilicus, on the under-surface. In both, the surface layers of the shell are often worn away near the apex.

A good low tide on southern and western shores is needed to expose the largest, and much the most handsome, of British species. *Calliostoma*

PLATE 19
a. BLUE-RAYED LIMPET, *Patina pellucida*, on fronds of the Tangle-weed, *Saccorhiza bulbosa*
b. PAINTED TOP-SHELL, *Calliostoma zizyphinum*

PLATE 19

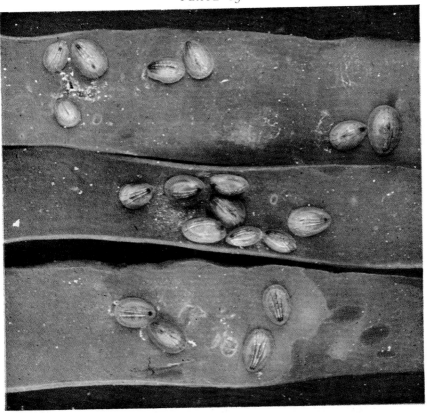

Photographs by D. P. Wilson

PLATE 20

D. P. Wilson

zizyphinum, the painted top-shell, cannot be mistaken. As shown in Plate 19b (p. 126), it is truly like a top with a flat base, up to an inch across, straight sides and a sharply pointed spire and comprises eight to ten closely applied coils or whorls compared with the six more rounded whorls in the other species. The shell is typically pink or yellow with streaks of red but white shells occur in some areas. When moving about it exposes, in addition to the usual pair of head tentacles with prominent eyes near the base, four tentacles on each side of the body above the foot. In the colder climate of our eastern coasts this species never comes above low tide level and top-shells in general are less hardy than periwinkles or limpets.

Most of the top-shells discharge their sexual products freely into the sea like limpets and chitons. But the painted top-shell, which spawns in the spring and summer, lays its eggs in a long gelatinous ribbon, fertilisation taking place after they are laid. The young hatch out as small crawling individuals so that the planktonic stage in the life history is omitted.

Hitherto we have been concerned with snails that browse upon the seaweed and lichens growing on the rocks, or else with barnacles that collect a more varied diet of microscopic plants and animals from the plankton. Now we come to consider a carnivorous snail, the dog-whelk, *Nucella* (*Purpura*) *lapillus*. This occupies much the same area as the common periwinkle, namely from about the middle of the shore down to low tidal levels, but it is not quite so widespread as that animal ; it does not extend on to such soft substrata or into estuaries. It is also absent on fully exposed shores where there are no protective crevices. But it remains among the commonest and most successful of shore animals.

The dog-whelk is about the same size as the common periwinkle but belongs to another group of snails and has very different habits. It is widely distributed from the coast of Portugal round the coasts of Europe, excluding the Baltic, to the Murmansk coast, along the west coast of Iceland and from the coast of Newfoundland south to the region of New York. As shown in Plate 22 (p. 129), the shell varies greatly in colour, it may be white, yellow, mauve-pink or brown-black. Possible reasons for this colour variation are discussed later. The shell

PLATE 20
FLAT PERIWINKLES, *Littorina littoralis*, on Serrated Wrack, *Fucus serratus*

also varies in form and those of animals living in the sub-littoral zone may be covered with wavy ridges. The allied *Thais lamellosa*, which fills a similar place in the economy of life on the Pacific coasts of North America, is even more variable in colour and form.

' The shell is thicker than that of the common periwinkle and if the two be compared an obvious difference between the shape of the opening will be seen. In the periwinkles this is smoothly rounded but in *Nucella* the under-side is drawn out and deeply grooved. When the animal is active a siphon composed of a strip of tissue curled round to form a tube is extruded•along this groove. Through it water enters the gill chamber. Such a device is typical of snails which have become adapted for life on soft substrata of mud or sand ; they can plough their way through this while extending the siphon above it where it can draw in clean water for respiration. The assumption is that in past time the ancestors of the dog-whelks moved back on to rocky surfaces and so on to the shore where rich food awaited them.

If we further compare the shells of periwinkles of different ages we find they differ only in size. In *Nucella*, on the other hand, there is a clear distinction between a young growing animal and an adult. Indeed, since the size the animals attain varies a good deal with the conditions of life, adult shells may be found which are much smaller than growing ones. In the latter the opening of the shell is wide and its outer surface smooth. But when the animal becomes sexually mature it ceases to increase the size of the shell ; instead the shell becomes thickened round the margin so that the opening is reduced while a series of tooth-like processes project in from its outer margin. Stoppage of growth at sexual maturity occurs only in certain snails, the small cowries are the only other good examples from our shores. The majority of other sea-snails continue to grow, with much-diminished speed it is true, for some time after they begin to spawn.

The dog-whelk, we have suggested, came late to rocky shores after adaptation to other habitats. This is also to be inferred from its feeding habits. In any new environment—and for a moment let us think of the shores as bare before their first colonisation by living things—the first occupants must be plants obtaining food from the

---

PLATE 21
GREY AND FLAT TOP-SHELLS, *Gibbula cineraria* and *G. umbilicalis*, with Common Limpets, *Patella vulgata*, on pebbles

PLATE 21

PLATE 22

John Armitage

inorganic constituents of water, soil and air. Then come herbivorous animals feeding upon them and only after them the carnivores which prey on their fellow animals. So we must assume a previous population of herbivorous animals before the dog-whelks appeared on the shore. Clearly it is not every animal that could prey on the molluscs and barnacles of the shore encased in stout shells and hardly to be dislodged by animals when they can resist the full force of the sea. The dog-whelk attacks insidiously. It has a lingual ribbon like other snails but it is narrow, with few and prominent teeth in each row, and carried at the end of a proboscis that can be protruded for some distance. When feeding the animal settles on to its prey and slowly bores a narrow hole through the shell. This is a purely mechanical process although there are snails, such as *Natica* (Fig. 72, p. 237), which aid the process with acid that dissolves the limy shell of the prey. Once the shell is perforated the proboscis extends still further into the tissues of the contained animal, which are rasped out by the teeth and passed back, as on a conveyor belt, into the throat.

Most of the browsing snails, including the common limpet, the flat and the grey top-shell and the common and the flat periwinkles, are preyed on by dog-whelks, but their principal food comes from two animals which cannot move away, however slowly, if attacked, namely barnacles and mussels. Dog-whelks feed on either *Balanus balanoides* or *Chthamalus stellatus* and for that purpose extend well up into their zone. The laborious process of boring is here unnecessary ; the valves are simply forced apart and the proboscis inserted into the tissues so exposed. It is difficult to determine exactly how the valves are separated because the barnacle is completely covered by the foot of the snail during the process. It has been suggested that the purple dye, purpurin, which is produced by the animal and is certainly highly poisonous, may first kill the barnacle causing the muscles to relax.

Where mussels are abundant they form the chief food. These bivalve molluscs still await description ; it is sufficient for the moment to say that they are frequently among the commonest animals attached to the surface of intertidal rocks. Older shells are bored by the dog-whelk but access to the bodies of young mussels is frequently obtained

---

PLATE 22

D O G - W H E L K S, *Nucella lapillus*, of varius colours on rocks with the Acorn-barnacles and Mussels on which they feed by forcing the two valves apart. And even a small mussel will provide more and richer food than a barnacle. Consequently dog-whelks grow larger on this diet than on barnacles. Moreover the colour of the shell is affected by it ; the brown-black and mauve-pink individuals (see Pl. 22, p. 129) being those that have fed primarily on mussels whereas a white shell indicates that barnacles have formed the food. An animal which has transferred from one food to the other reveals this in an abrupt change of colour in the shell. The yellow colour, it may be mentioned, appears to be associated not with diet but with exposure to wave action although the underlying reason for this remains obscure.

Normally dog-whelks continue feeding on barnacles or mussels and only change their diet if the supply runs short. Thus a colony of mussels usually has a good chance of establishing itself while the snails are feeding on the surrounding barnacles. When the change is made, owing to lack of barnacles, the dog-whelks seem uncertain at first how to tackle their new food. They are described as sometimes boring through empty shells, or settling within empty ones and boring outwards. But gradually they become adjusted in their reactions and proceed about the business efficiently. In due course the supply of mussels may in its turn give out. But time has been allowed for the re-establishment of the barnacles and so the dog-whelks are able to return to their original food.

The animals have to pair before eggs can be laid and for this purpose they collect in large numbers together, usually in crevices, where they may also go for protection in stormy weather or in the colder months of the year. As with the periwinkles, a protective case is laid down around the eggs but this case is then secured to the surface of the rock. Throughout most of the year, but especially in winter and spring, large numbers of straw-coloured capsules, each about the size and appearance of a grain of corn, will be found sticking upright on the rock surface, usually in crevices or on the under-side of boulders (Pl. XVIb, p. 119). The animal is said to take about one hour to produce each one, anything from six to thirty-one being laid at any one time, although, in repeated spawnings, a single snail can lay between two and three hundred. In each capsule are some hundreds of yolky eggs but the great majority of these are not fertilised and serve only for the nourishment of the few that develop. After about four months these emerge, some ten or twelve in number, as small, fully shelled individuals (Fig. 37, p. 131). Young ones are found in greatest

numbers well below the zone occupied by the adults and it seems
probable that they are carried
down the shore by the retreating
tide or in the undertow of wave
action. Here they find themselves
in a region where the stones are
covered by the coiled, limy shells
of the small tube-worm, *Spirorbis
borealis* (Pl. XVIIIa, p. 139). On
these they apparently feed until
about one-third of an inch high
when they migrate up the shore to
the barnacle zone. The entire life-
history is thus spent on the shore on
which the animals are also completely dependent for food.

The adaptations of the dog-
whelk to shore life are not obvious.
There is the strong shell, it is true,
which permits much rolling about
by the waves. There is an operculum but it is seldom employed to
close the shell except when the
animal loses its grip on the rock.
When picked off exposed rocks,
they are usually found to be adhering by the foot but this has

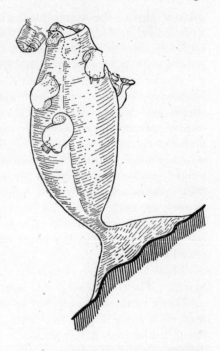

FIG. 37—Egg-capsule of dog-whelk,
Nucella (Purpura) lapillus, attached to rock
surface and with young animals emerging
from it after completion of development.
Ten times natural size. (Modified after
Ankel, *Verhandl. der Deutschen Zoolog.
Gesellschaft*, 1937, p. 77.)

none of the wide adhesive grip of a limpet. To avoid desiccation they
do no more than pull the shell down against the surface of the rock.
But the tissues are thick and tough and probably retain water better
than do those of periwinkles. In any case, the dog-whelk is a successful
shore-living animal.

The common mussel (Pl. 31, p. 166) is widely distributed along
the coasts of Europe and, like the dog-whelk, also inhabits the north
Atlantic shores of America. No shore animal grows in such dense
masses or can establish itself so quickly (Pl. XVII, p. 138). In the
campaign which freed the Netherlands from German invasion in 1944,
the dykes guarding the island of Walcheren were breached and the

greater part of the island flooded with sea water. When the breaches were closed and the water pumped out about a year later the surface of roads and the sides of houses and fences were found covered with mussels, which even hung in clusters like fruit from the branches of trees.

The bivalve molluscs, as we have already had occasion to note, are primitively inhabitants of soft substrata and we shall find them in varied and rich abundance in sand and mud. The broad foot of the snail is an admirable organ for slow progression over a rocky surface or for attachment to this. The laterally flattened foot of a bivalve is as well fitted for movement through sand or mud but not for gripping a hard surface. Hence those bivalves that live between tide marks on rocky shores are attached in other ways, either by cementation as in oysters, or, more usually, by means of tough threads which constitute the *byssus*. In the former case the animal is incapable of movement, but not necessarily in the latter where the foot is not lost although it may be reduced.

Mussels are the commonest of bivalves that are attached by a byssus. The threads of this are produced by a gland within the substance of the foot and they issue from it in the form of a thick fluid that runs along a groove which extends down the hinder surface of the thin and very extensive foot. This organ plants the threads. It is directed first in one direction, then in another, the sticky fluid running along the groove and spreading out into a rounded disk of firm attachment where it meets the rock (Fig. 38, p. 133). Almost immediately each hardens into a tough thread. Finally the mussel is attached by a diverging mass of threads like the guy ropes of a tent. In whatever direction the sea strikes, some of the threads are in a position to take the strain. The shape of the mussel helps in resisting wave action. It is narrow at the anterior end and broad at the other. The threads are directed forwards so that the animal tends to swing round with the narrow end facing the force of the sea, although in a dense mussel bed the animals are so tightly packed that movement is impossible. If the threads are broken new ones are readily formed and the animal may move about for a short time by means of the foot before it re-attaches itself. Moreover it can, by this means, raise itself above mud and sediment. Mussels will often be found on the shores of muddy estuaries. They are not attached to the mud but to stones embedded in this and if more mud descends on them they can cast off the old attachment and make new and longer threads that enable them to remain on the surface. The byssus apparatus, which is probably a modification of the glands used in the snails for lubricating the sole of the foot with mucus, provides an admirable means of attachment, tough yet yielding and capable of repair and modification with changing circumstances.

The form of the mussel is not typical of bivalves. The anterior end is reduced and the posterior enlarged, and inspection within will reveal the presence of a large posterior adductor (drawing together) muscle and a small anterior one. In a typical bivalve, such as a cockle, the animal is more symmetrical and the two muscles of about equal size. The simplest bivalves to possess a byssus are symmetrical, with the hinge in the middle instead of near the anterior end as it is in the mussel, and the byssus passing straight down in the middle region between the two valves. We have only small representatives of such bivalves, species of ark-shell, *Arca*, which may be found with some difficulty attached in the shelter of rocks near low tide level on the south coast. The asymmetry of the mussel has been attributed to its exposed habitat, or, as we may better express it, such a change in

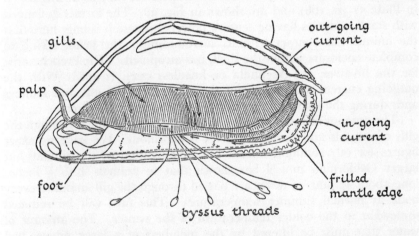

FIG. 38—Common mussel, *Mytilus edulis*, view of the mantle cavity after removal of the left shell valve and mantle fold. Water is drawn in at the hind end below the gills, is strained through these and passes out above them, also at the hind end. Arrows on the gills show the direction of passage of food particles towards the palps which guard the mouth. Other, broken, arrows indicate passage of excess particles back to the position of entrance where they are from time to time extruded by contractions of the adductor muscles. (After Orton, *Journal Marine Biological Association*, Vol. IX, p. 459.)

form was advantageous to its possessors, which were consequently gradually selected in that direction in the competition for existence on the shore.

Like the great majority of bivalves, mussels feed on the plant plankton. This is carried out by the gills, enormously enlarged in connection with their added function of straining this food from the water. As shown in Fig. 38 (p. 133), they extend along the entire length of the body and each consists of two broad plates composed of great numbers of parallel filaments. Water passes between the latter, leaving the food particles on the surface of the gills. One set of ciliary hairs causes the water current, other sets carry the food, after it becomes entangled in mucus, to the mouth, which is guarded by pairs of ridged flaps known as palps. These also are richly ciliated and have a regulative function, preventing too large or too frequent masses from passing to the small mouth which would easily be blocked. Water enters the gill cavity at the broad, hinder end of the body by way of an ingoing opening below the line of the gills, and leaves by an outgoing opening above this. Both of these apertures may be distinguished in Plate 31 (p. 166) and are shown in Fig. 38. The former is fringed with small tentacles having sensory powers because it is only here that the animal has contact with its surroundings ; the head, owing to complete enclosure within the shell, has atrophied. (The French name for the bivalves, the Acephala or headless ones, is apt.) With the outgoing current are passed out the waste from the gut and kidneys and, during the spawning period, the reproductive products.

The amount of water that is continually being strained through the gills of mussels and similar bivalves is enormous. We have no exact figures for our common mussel but investigation of the similar but larger Californian mussel has shown that in animals 4 to 5 inches long about $4\frac{1}{2}$ pints of water are passed through the gill chamber every hour at normal summer temperatures. This rate will be reduced somewhat in the colder temperature of the winter. The amount of water that must be filtered by the members of a large mussel bed composed of tens or hundreds of thousands of individuals can be imagined, also the immeasurable numbers of microscopic plants of the plankton which must be collected in this way.

Mussels are dependent on the sea for reproduction as well as for food. They spawn in the early months of the year. Suitable temperature provides the initial stimulus but the presence of egg or sperm in the water quickly excites all individuals to spawn so that the water above the mussel bed soon becomes alive with sexual products and fertilisation is assured. A period of larval life in the plankton is followed by settlement but the young are at first active and move upshore by means of the foot before they attach themselves, more or less permanently, by the byssus. Young ones placed in a jar of water will pull themselves up the side of this by anchoring with a few strands of byssus, then extending the foot higher and planting new threads, detaching the old threads and pulling themselves up on the new ones. So the process goes on till the animal reaches the surface of the water. Speed of growth varies very greatly, dependent on food and also, in estuaries where mussels live readily, on salinity ; much lowering of this stunts the growth. Under ideal conditions they increase in length by about one inch annually for the first two or three years after which growth slows down. But such conditions are rare in nature, competition is too intense, and on a mussel bed it probably takes five to seven years for a mussel to grow to a length of three inches.

Mussels extend over rocky shores and on muddy shores where there are stones for attachment, they pass for some way up estuaries and far into the steadily diminishing salinity of the Baltic Sea. They occur freely in the sub-littoral zone. Only much exposure to wave action limits their distribution and here they yield place to the barnacles and some of the intertidal snails.

The best-known bivalve to attach itself by cementation is the native oyster, *Ostrea edulis*, but this is not common between tide marks and indeed all too rare these days in the shallow depths of tidal creeks and estuaries which are its usual habitat. In this bivalve the foot is present for only a brief period immediately after settlement while it is searching for a suitable place for attachment. When this is found, cement is poured out from the modified byssus gland and the animal comes to permanent rest with the left shell valve fixed by this calcareous cement to the rock. The internal structure has been modified still further than that of the mussel because, in addition to the loss of the foot, only one muscle, the original posterior adductor, is present and has come to lie in the centre. The same is true of the scallops, which we shall be discussing in more detail later.

The saddle-oyster or silver shell, *Anomia ephippium*, is not uncommon on the sides of or under stones near low tide mark. The fragile white shell is closely adpressed to the surface of the rock, the curvature of which it follows. It can easily be confused with the true oyster but if the upper valve is removed striking differences in structure are revealed. The under-valve has a large perforation giving passage to the byssus, which is hard, owing to calcification, and forms a compact mass, round in cross-section. Here also only one adductor muscle is present. The saddle-oysters are strange creatures, more closely allied to the mussels than to the oysters but with many unique features. They may be said to have solved the problem of resistance to wave action by their close application to the rock surface, but not the problem of exposure to the air because they do not occur high on the shore and even within their limits of distribution upon it are usually well sheltered.

PLATE 23
CLUB-HEADED HYDROID, *Clava squamata*, on Knotted Wrack. *Ascophyllum nodosum*

PLATE 23

D. P. Wilson

PLATE 24

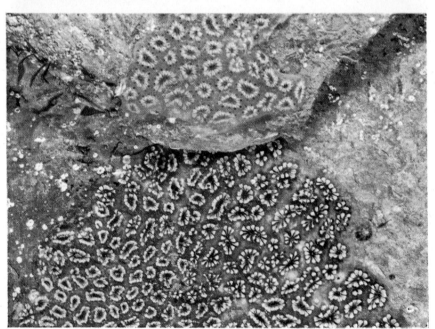

Photographs by D. P. Wilson

CHAPTER 10

THE INTIMATE FAUNA OF
WEED AND ROCK

"Ruder heads stand amazed at those prodigious pieces of Nature, Whales, Elephants, Dromidaries and Camels ; these, I confess, are the Colossus and majestick pieces of her hand : but in these narrow Engines there is more curious Mathematicks ; and the civility of these little Citizens more neatly sets forth the Wisdom of their Maker."

SIR THOMAS BROWNE : *Religio Medici*

FOR every species of animal that can withstand the rigours of full exposure on a rocky shore there are hundreds that dwell protected amidst seaweed or between and under rocks. There they are sheltered both from the full force of the waves and from exposure to a dry atmosphere ; from the excessive heat of summer or cold of winter. Anything like a full description of the rich fauna which diligent search will reveal is out of the question ; we can only take the more conspicuous species as typical of all and by a general survey of life under sheltered conditions fill in some of the remaining details of our picture of life on a rocky shore.

The shore weeds provide a safe environment for a vast population of animals and also give food to many. Estimates have been made by Mr. J. Colman of the numbers of animals which inhabit the commoner types of intertidal algae on the south coast of Devonshire, and the results are striking. They reveal a greater density of population than that of the richest soil. We will consider first the weed that grows highest on the shore, the channelled wrack, *Pelvetia canaliculata*. In this region, which is so frequently exposed, the animals must be resistant to desiccation and be able to feed largely on the weed or on their fellows or scavenge omnivorously. Hence the fauna is relatively sparse

PLATE 24

a. OATEN-PIPES HYDROID, *Tubularia larynx*

b. GOLDEN-STARS SEA-SQUIRT, *Botryllus schlosseri* ; two colonies spreading over the surface of a rock

but nevertheless consists of some 2,400 animals per square metre of rock surface. The more obvious animals consist of periwinkles, especially the rough species, *L. rudis*, and two crustaceans, the amphipod, *Hyale nilssoni*, and the isopod, *Ligia oceanica*. The former is a "hopper" and also occurs in *Enteromorpha* pools high on the shore and in brackish waters generally. It is thus an extremely hardy animal and, like its fellow amphipods, scavenges on debris of all kinds. The sea-slater, *Ligia* (Fig. 8, p. 28), is one of the commonest of shore animals and lives in crevices in rocks or in sea walls about high tide mark. Although young animals may be found in the daytime lurking among the fronds of *Pelvetia*, the sea-slater is nocturnal and emerges in great numbers after nightfall when the tide is out. Then it descends to the shore to feed on *Pelvetia* and on the lower fucoid algae although any type of vegetation with animal debris of all kinds is freely taken. The animals are very sensitive to light and scuttle into shelter when the beam of a torch is cast upon them. Even the light of the moon is sufficient to keep them under cover and where thousands may be out foraging on a dark night only a few dozen are discovered in bright moonlight. Avoidance of light is connected not only with the danger of desiccation in daytime but also with the threat of attack from sea-birds and even shore-crabs, because *Ligia* is a large isopod, broad in body and about one inch long, and so easily catches the attention of marauders.

The flat wrack, *Fucus spiralis*, carries rather more than double the population of animals over a similar area and this fauna is more varied than that on *Pelvetia*. The hopper, *Hyale*, continues abundant but young slaters do not, while periwinkles are represented by a few of the flat species. The major increase is in the numbers of minute copepod crustacea, certain types of which abound on all fucoids. The short tufts of the common lichen, *Lichina pygmaea*, which extend from high water level to about the middle of the shore, support an amazing fauna of over a quarter of a million animals per square metre. There are various, mainly young, periwinkles, small isopod crustaceans, many insect larvae and mites, all of them probably feeding mainly on the lichen, but especially numerous is the minute bivalve mollusc, *Lasaea rubra*. This, of course, takes its food from the sea, which covers it long enough to permit it to strain adequate supplies of plankton. Such is the abundance of this bivalve that, despite its small size, it demands a short description. It grows to lengths of a little over one millimetre

PLATE XVII

H. H. Goodchild

a. General view of mussel beds at Conway, North Wales

D. P. Wilson

b. Closer view of Mussels, *Mytilus edulis*, exposed at low tide ($\times\frac{1}{2}$)

PLATE XVIII

D. P. Wilson
a. Coiled Serpulid Tube-worm, *Spirorbis borealis*, growing on Serrated Wrack (×½)

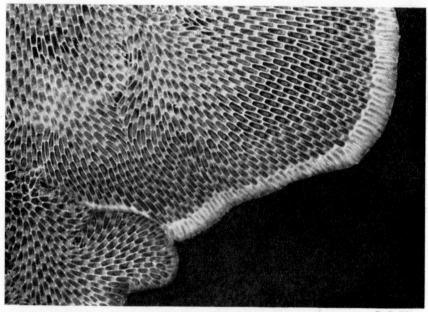

D. P. Wilson
b. Polyzoan, *Membranipora membranacea*, growing over a frond of *Laminaria*
Photographed under water (×6)

and has a reddish-brown shell. Like the mussels, it is attached by byssus threads, sometimes to lichens but often within the shelter of empty barnacle shells. It can move about by means of the slender foot which, when extended, may exceed the length of the shell. The mode of development is of especial interest. The males shed their sperm into the sea but the females retain the large, yolky eggs within the gill chamber, where they are fertilised by sperm carried in by the water currents. Hence development occurs within the shelter of the gill chamber and, like those of the rough periwinkle, the young emerge at a stage when they can immediately colonise the shore. This habit of incubation of the young, unusual among bivalves, is probably the prime reason for the great success of these small animals.

Compared with the lichen zone, the area covered by the bladder-wrack, *F. vesiculosus*, carries but a limited fauna of under fifty thousand animals for the standard area. Minute copepods and mites which can hold on in the rough seas to which this weed is often exposed, form a large proportion of these. But the longer fronds of the knotted wrack, *Ascophyllum nodosum*, with its tufts of red *Polysiphonia*, bear a much richer population of over two hundred thousand animals. The greater shelter provided by these weeds and the calmer water in which they grow probably account for this denser population. Here for the first time will be found great numbers of attached or encrusting animals, especially small hydroids. Among the commonest, and the easiest to see because of its pink colour, is the beautiful little *Clava squamata* (Pl. 23, p. 136). This is an example of a naked hydroid in which the small feeding polyps are not enclosed in a skeletal cup. It grows in little clusters of simple stems arising from a common base. An example of the other type of hydroid in which the polyps can contract within the shelter of the skeleton is provided on the same weed by the common sea-oak, *Dynamena* (*Sertularia*) *pumila* (Fig. 39, p. 140). The colony is much branched and grows in little tree-like tufts to heights of about one inch. A series of such tufts arise from ramifying basal growths. As in all hydroids of this type, the skeleton is tough and may long persist after the death of the colony.

These hydroids can only be adequately viewed after they have been placed in sea water, when the polyps gradually emerge and the delicate tentacles expand in a ring around them. The majority of British hydroids occur in the sub-littoral zone but there are a number, apart from the two mentioned above, which are characteristically shore

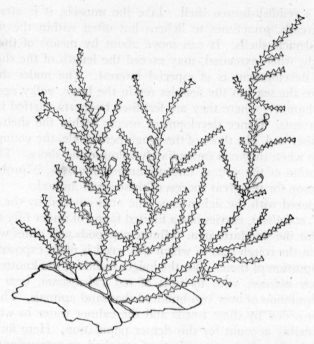

FIG. 39—Sea-oak, *Dynamena* (*Sertularia*) *pumila*, showing branched colony bearing occasional capsules in which the eggs develop, there being no free medusae. Twice natural size. (From Hincks, *British Hydroid Zoophytes*, Plate LIII, fig. 1.)

forms living in damp places attached to weed or rocks or, as we have already seen, inhabiting rock pools. We may most conveniently mention here the relatively large *Tubularia*, a genus of naked hydroids which is well worth seeking at low water of spring tides when specimens may be found at the base of rocks, sometimes half smothered with mud. There are two common species, *T. larynx*, shown in Plate 24a (p. 137), which has branching stems, and the somewhat larger *T. indivisa* with longer simple stems. The pink-coloured polyp is large with two rings of tentacles, one around the mouth and the second lower down where the polyp widens considerably. Between them cluster the medusoids, like little bunches of grapes. They never leave the parent hydroid and the young are liberated as larvae already possessing the first ring of tentacles. They soon settle and establish a new colony. John Ellis described these animals in his *Essay Towards a Natural History of the*

Corallines in 1755 where, referring to their stems, he writes: "These Tubes in the dried specimens have the Resemblence of Oaten Pipes; that is, Part of an Oat-straw, with the Joints cut off."

Many of the littoral hydroids have in some measure freed themselves from dependence on the sea because the medusoid generation is reduced, as in *Tubularia*, or absent, as in *Dynamena*. The alternation of fixed and free-living generations has already been noted as a primitive feature in these coelenterates and it persists in the hydroids that live on *Laminaria* and elsewhere in the sub-littoral zone. The sexual products which are formed and liberated by the medusae are, in shore hydroids, often retained on the fixed generation. Hence the period spent free in the sea is reduced to a short larval life, with a consequently greater chance of settlement on the shore.

A dense brown encrustation frequently found on the stems of *Ascophyllum* and other fucoids is formed by colonies of the polyzoan sea-mat, *Flustrella hispida* (Fig. 40). The surface of the skeleton is studded with many horny spines which protect the animals after these have withdrawn each into its particular cavity. When expanded their tentacles cover the skeleton with a semi-transparent milk-white or bluish film. The tiny coiled and limy tubes of *Spirorbis borealis* (Pl. XVIIIa, p. 139) are unmistakable but resemble the habitation of a snail rather than of a worm and this likeness is increased after inspection with a lens which reveals that the opening is closed by an operculum. This animal is the commonest representative of the serpulid tube-worms on our shores. All such worms are distinguished by the possession of a stout limy shell and the modification of one of the tentacles to form an opercular plug, as shown in Fig. 4 (p. 24). The remaining tentacles expand to form a crown around the pro-

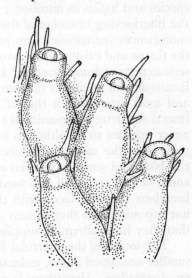

FIG. 40—Polyzoan, *Flustrella hispida*, showing compartments or zooecia each with protective spines around the openings. Greatly enlarged. (From Hincks, *British Marine Polyzoa*.)

truded head and they collect food from the water by a process of filtration, employing groups of cilia like those on the gills of a bivalve mollusc. Such worms appear ideally suited for life on the shore, protected by a stout shell which is firmly cemented to weed or rock and with an operculum to prevent desiccation. *Spirorbis* is very common and so are two other species we shall encounter later on rocks, but all are most numerous on the lower levels of the shore.

Of animals which crawl over the weed, the largest are flat periwinkles which are found on the two kinds of bladdered wrack and on *F. serratus* (Pl. 20, p. 127), the amphipod, *Hyale*, which ranges from the *Pelvetia* zone as far as the *Ascophyllum* zone and even lower, together with an isopod, *Idothea granulosa*. This is a smaller and more graceful animal than *Ligia*, about half an inch long and relatively narrow. It has much the same habits as the larger species, feeding voraciously on all that comes to hand and most frequently by night, but, as befits its lower zone on the shore, it is an active swimmer. The minute crustaceans, mites and worms inhabiting the weed are diverse in species and legion in numbers ; many of them find ideal homes among the interlocking branches of the clusters of red *Polysiphonia*. One small roundworm (nematode) lives parasitically on *Ascophyllum*, penetrating the tissues and raising warty patches on the surface. Similar worms are serious pests of domestic crops. Insect larvae form a significant although inconspicuous section of the fauna of the weed. Among the commonest and most interesting is that of the midge, *Clunio marinus*. The adult insects are true shore-animals ; the female is wingless and the male never appears to fly although it may use its wings as sails when skimming over the surface of pools. Insects have invaded the shore from the land and some have even established themselves in the *Laminaria* zone, but all are small and need patience and skill to discover. As air breathers they are faced with the problem of surviving submergence, not exposure, and they possess a variety of devices for trapping air so that they have adequate supplies of oxygen while covered by the sea.

The lowest of the intertidal fucoids, the serrated wrack, has a much smaller associated fauna, estimated at no more than one-eighth of that on *Ascophyllum*. Here there is a greater variety of snails, top-shells as well as periwinkles, with young individuals of the blue-spotted limpet, *Patina*. The polyzoans *Flustrella* and *Membranipora pilosa* are common, the latter forming a whitish mass on the stems. Both of these encrusting sea-mats are still commoner on the *Gigartina* and *Chondrus* which may

PLATE XIX

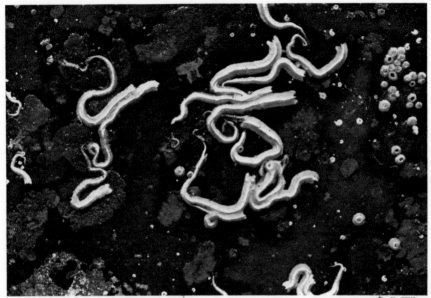

D. P. Wilson

a. Limy tubes of the Serpulid Tube-worm, *Pomatoceros triqueter* (×⅔)

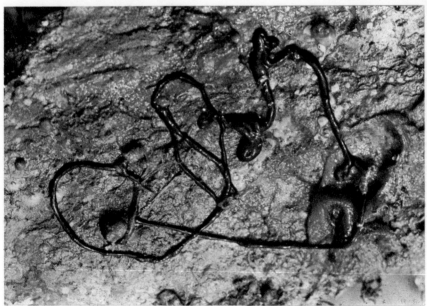

M. A. Wilson

b. Bootlace Worm, *Lineus longissimus*, photographed when exposed (×⅔)

PLATE XX

a. Scale Worm, *Halosydna gelatinosa*. Photographed under water (×1⅓)

b. Portion of a colony of the reef-building Tube-worm, *Sabellaria alveolata*, exposed by the tide (×¼)

grow as a thick turf on flattish rocks (Pl. VI, p. 69) overlapping the *Ascophyllum* and *F. serratus* zones. These red seaweeds provide close shelter with their stiffly growing fronds and a third polyzoan, *Alcyonidium hirsutum*, is frequent upon them. This is distinguished from the others by the fleshy and lobed nature of the colonies, which are grey in colour. A number of isopod and amphipod crustaceans with many wandering worms also find shelter and food among these weeds. Area for area, *Gigartina* has a richer fauna than any other weed except *Ascophyllum* and *Lichina*.

The broad fronds of *Laminaria* and allied tangles exposed at low water of spring tides frequently bear beautiful growths of hydroids and polyzoans. The commonest of the former is *Obelia geniculata* which may cover wide areas like a miniature forest of little zigzag stems. Each little side branch bears a terminal polyp and at the angle between these branches and the main stem may arise elongate receptacles in which the swimming bells or medusae are formed. They finally emerge through an apical opening, when a sudden contraction of the bell frees them from the parent hydroid. Broad patches of the polyzoan sea-mat, *Membranipora membranacea*, are unmistakable. They form thin films but with each chamber clearly visible, thus giving a lace-like texture to the colony which is bounded by delicately curved margins of active growth (Pl. XVIIIb, p. 139). The colony appears as though in process of slowly flowing over the flat surface of the weed. Small snails are not uncommon here, not only the blue-spotted limpet, *Patina pellucida*, but the beautifully banded *Lacuna vincta*, like a smaller, rather more pointed periwinkle. This lays little rings of white spawn easily seen against the dark-brown background.

The holdfasts of these large weeds with the ample protection provided by their over-arching struts provide the richest collecting area. On the same basis of area, this region alone of these tangle-weeds supports an estimated population little less than that of *Gigartina*. It is largely a different population comprising many animals not found higher on the shore. The taller variety of *Patina pellucida*, described in the previous chapter, occurs here and with it are a number of other snails with young specimens of both the common and the horse-mussel and also saddle-oysters and other bivalves. There is a wealth of encrusting organisms, sponges and compound sea-squirts as well as hydroids, polyzoans and barnacles. A wide variety of crustaceans find shelter in these holdfasts—microscopic copepods and larger amphipods and isopods with occasional crabs. The greatest wealth of all is in worms, especially the segmented bristle-worms. All manner of these dwell here, errant forms that crawl or swim and sedentary forms inhabiting tubes of lime, like the serpulid *Spirorbis* and the larger *Pomatoceros* (Pl. XIXa, p. 142), or of parchment-like consistency, like the sabellids, or of agglutinated sand grains like the terebellids that we shall encounter in hosts on sandy shores. To the specialist collector there is no more profitable ground ; to the amateur observer no better revelation of the richness and diversity of animal life on the fringe of sea and shore.

The true wealth of the shore is only fully revealed to those who search diligently. We have already peered into rock pools and searched through masses of intertidal weed and examined the obscurities of the holdfasts of tangle-weeds. Now we must be prepared to turn aside armfuls of weed to expose the boulders beneath and, where there are great rocks with tortuous passage between, to bend cautious way between these until we can look up at their overhanging surfaces. It is under such conditions of damp shade from weed or from the wide overhang of rocks that we can get the best view of encrusting animals that cannot withstand the full exposure to which barnacles, periwinkles, mussels and the like are adapted.

Under ideal conditions the surface of the rock will be covered with many-coloured masses of sponges, compound sea-squirts and polyzoans. No three groups of marine invertebrates could be more different in fundamental structure, but in their manner of growth and in their dependence on the sea water for food and reproduction they have much in common. And none can withstand exposure by the retreating tide except where there is little danger of desiccation.

Wide areas of sheltered surface are often covered with a dense fleshy encrustation, most usually green but on occasion yellow or brown, of which the only structural feature consists of scattered rounded

---

PLATE 25 *

a. ~~Crumb-of-Bread Sponge, *Halichondria panicea*, growing in crevices of an exposed rock otherwise encrusted with the red coralline weed, *Lithothamnion*~~

b. Sponge, *Halichondria bowerbanki*, Simple Sea-Squirt, *Ascidiella aspersa* and Colonial Sea-Squirt, *Morchellium argus*, all characteristic of very sheltered waters, exposed at extreme low water of spring tides at the Salstone, Salcombe estuary

* 25a. Sponge, Hymeniacidon sanguinea, growing on a stone upon a muddy estuarine shore.

25b Crumb-of-Bread Sponge, Halichondria panicea, growing in crevices of an exposed rock otherwise encrusted with the red coralline weed, Lithothamnion

PLATE 25

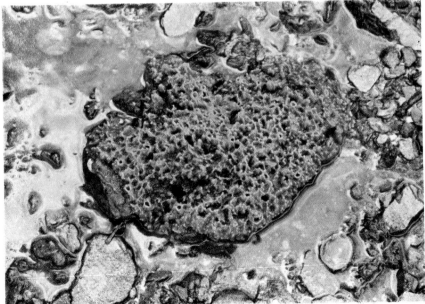

Photographs by D. P. Wilson

PLATE 26

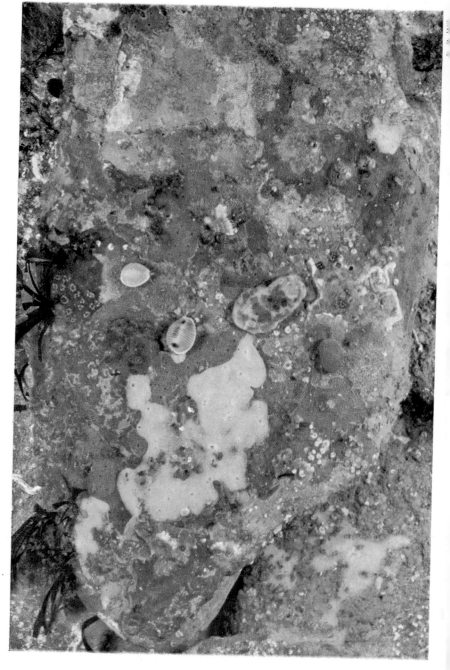

openings at the summits of small prominences like the craters of miniature volcanoes. This is the common crumb-of-bread-sponge, *Halichondria panicea* (Pls. 25a and 26), and the openings are oscula, such as those that Grant saw vomiting forth an " impetuous current " and knew then that he was observing an animal and not a plant. The openings through which water is drawn into the honeycomb of canals which ramify through the mass of the sponge are too minute to see. Although a true animal, the growth of the sponge is similar to that of a plant and equally influenced by external factors, especially by water movements. In general the quieter the water the thicker will be the encrusting mass. On the base of seaweeds it may be found growing as rounded lumps. The texture is unmistakably spongy ; when pressed water is forced out but on release of pressure the tissues swell again by drawing in water from those around them. There is little cohesion and the sponge can easily be pulled apart when the small silica spicules which form the skeleton can be felt if not seen.

Similar masses of blood-red, scarlet or orange which are also common will most probably be *Hymeniacidon sanguinea* (Pls. 38a and 40, pp. 245, 253) which, apart from its colour, is distinguished from the crumb-of-bread-sponge by the smaller and more numerous oscula. *Ophlitaspongia seriata*, shown in Plate 26, is also red but somewhat duller in shade and of a smoother texture. A much simpler type of sponge, very different in growth form, may be found attached by a short stalk and hanging down on the under-side of ledges. The purse-sponge, *Grantia compressa*, named after Grant, is flattened and so the better able to withstand exposure when it lies against the rock surface. It is an inch or more long with a single large osculum at the free end and a large internal cavity into which water is drawn through numerous minute pores. It belongs to a small group of relatively simply con-structed sponges in which the spicules are of lime instead of silica. A cylindrical sponge of the same type, *Sycon coronatum*, may sometimes be

PLATE 26
UNDER-SURFACE OF A LEDGE OF ROCK showing encrusting Sponges, *Halichondria panicea* (yellow), *Ophlitaspongia seriata* (red), and *Microciona atrasanguinea* (dark red), Compound Sea-squirt, *Botryllus*, and Polyzoan, *Umbonula verrucosa* (rose-red) with browsing Cowries, *Trivia arctica* and *T. monacha*, Sea-slugs, *Archidoris britannica* (mottled) and *Rostanga rufescens* (red) and Snail, *Nassarius incrassatus*

found under similar conditions, but is not usually exposed for so long.

Mention has already been made of various polyzoans that occur under rocks as well as on weed but there are a number of encrusting species which are specially characteristic of rock surfaces and in which the skeleton is impregnated with lime. One of these, which forms thin rose-red sheets, *Umbonula verrucosa*, is shown in Plate 26 (p. 145). All sea-squirts are attached and the compound forms are amongst the commonest of encrusting organisms. The most ubiquitous is the golden stars sea-squirt, *Botryllus schlosseri* (Pl. 24b, p. 137). The mass of the colony is thin and gelatinous and everywhere unmistakably dotted with star-like groups of individual animals each one drawing in water with contained food and oxygen through its own particular aperture but all members of each " star " having a common opening for its extrusion. The range of colour is very wide : the gelatinous matrix may be yellow, green, violet or blue with the individuals picked out in yellow or red. *Botrylloides leachii* forms similar thinly spread colonies but here the individuals are more irregularly grouped in long ovals that may branch and unite with one another in the common mass of jelly. No less common, but less conspicuous, are the grey masses of *Didemnum gelatinosum* in which the individual animals appear as scattered white dots, or the tougher *Trididemnum tenerum* with the matrix strengthened with limy spicules. All of these animals may be found on weed as well as on rock.

With these closely adherent masses of sponge, polyzoans and compound sea-squirts, are pendent animals such as the purse-sponges, solitary, simple sea-squirts and also sea-anemones, not uncommonly the plumose species, *Metridium senile* (Pl. 9b, p. 84), hanging down with the free end dilated by water retained by the closed mouth. All of these immobile animals are browsed upon by a variety of slow-moving carnivores, largely molluscs. A number of these are shown in Plate 26, including the sea-slugs *Archidoris britannica* and *Rostanga rufescens*, which feed on sponges, the carnivorous snail *Nassarius* (*Nassa*) *incrassatus*, and the little cowries, *Trivia arctica* and *T. monacha*. The latter are of interest as the sole British representatives of one of the largest and most conspicuous groups of tropical marine snails some of which attain lengths of several inches with a very massive and handsomely coloured shell. The cowrie shape is acquired only in adult life ; the shell of the growing animal has the usual spiral form, but gradually the outer whorl grows over the inner ones until the opening becomes elongated in the long axis of the animal. When active the animal extrudes not only the head and foot but also the mantle tissues which extend over the shell on either side and meet in the middle line above so that it is entirely covered. In this manner the shell, though it cannot grow further once the adult form is reached, is protected. The glossy surface which cowrie shells retain throughout the life of the animal, in striking contrast to the worn and pitted shells of many old snails, is due to this continual protection by the tissues. The small cowries are always found associated with compound sea-squirts. They feed upon them by means of a long proboscis, the lingual ribbon at the tip of which rapidly eats out the small animals from the containing jelly. In the breeding season the cowries make a small hole in the gelatinous mass and then deposit in this a vase-shaped capsule (Fig. 41) only a few millimetres high but containing several hundred small eggs of a bright-orange colour. After hatching the young larvae break their way through the mouth-like opening of the capsule and proceed to temporary life in the plankton.

The catalogue of what may be found on the damp overhanging surfaces of rocks is very far from complete ; there is abundance of worms, crustaceans and other molluscs crawling over and between the attached animals. No more can be done here than to draw attention to this rich environment and proceed, in a new chapter, to a survey of the still more varied fauna found between, and especially under, rocks and boulders, frequently further protected above by a thick carpet of fucoid weeds.

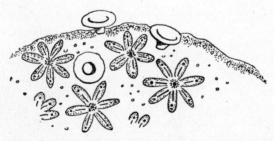

FIG. 41—Projecting lips of egg-capsules of the cowrie, *Trivia monacha*, embedded within a colony of the compound sea-squirt, *Botryllus*. Four times natural size. (From Lebour, *Journal Marine Biological Association*, Vol. XVII, p. 820.)

CHAPTER II

LIFE UNDER SEAWEEDS AND BOULDERS

"Dissected the Sea Urchin (*Echinus esculentus*). Very excited over my first view of Aristotle's Lantern. These complicated pieces of animal mechanism never smell of musty age—after aeons of evolution. When I open a Sea Urchin and see the Lantern, or dissect a Lamprey and cast eyes on the branchial basket, such structures strike me as being as finished and exquisite as if they had just a moment before been tossed me fresh from the hands of the Creator. They are fresh, young, they smell *new*."

W. N. P. Barbellion : *The Journal of a Disappointed Man*

MANY of the larger members of the shore fauna are seldom freely exposed when the tide is out. They have to be sought in narrow crannies between or under rocks or hidden among the gravel, sand or mud which is un-covered when loose boulders are turned (to be replaced later in their former position by all who value the amenity of the shore). There is protection in such localities from wave action and from desiccation and also a wealth of food in the form of fragments of vegetable and animal matter.

FIG. 42—Aristotle's Lantern. (From Forbes, *History of British Starfishes.*)

Worms of all kinds abound. The majority are segmented bristle-worms but others, as long or longer, with smooth and very slimy bodies belong to the unsegmented group of the ribbon-worms or nemertines. The most beautifully coloured of these worms is *Tubulanus annulatus*, shown in Plate 27a (p. 152), which lives within a tube of mucus under stones or in crevices on the lowest levels of the shore. It is excessively narrow but may be several feet long. A shorter worm, not usually more than three inches in length, is the pink ribbon-worm, *Amphiporus lactifloreus*, sometimes white or grey in colour and possessing a proboscis as long as the body and armed with a stylet. The commonest nemertine is probably the red-line-worm, *Lineus ruber*, usually of a reddish brown and three to six inches long. Like all such worms it is carnivorous ; it feeds on bristle-worms, which it engulfs whole like a snake, but with the aid of the proboscis. The allied *L. longissimus* (Pl. XIXb, p. 142), the bootlace-worm, sometimes dignified with the title of sea-snake, is much the largest of British nemertines. It is commoner in the shallow water of the sub-littoral zone than on the shore proper but may be found at low water of spring tides in the branched holdfasts of *Laminaria* where it intertwines into an elaborate tangle that is most difficult to unravel without breaking the soft body. Fully extended these worms may measure up to five yards, but they are very apt to contract into tight knots when handled, as shown in the Plate.

The bristle-worms are not restricted to a carnivorous diet like the ribbon-worms, nor to slow crawling motion. They are among the most successful of marine invertebrates and display a range of form and habit only equalled by the molluscs and crustaceans. Most typical in form and usually among the commonest are the ragworms which are often collected for bait. They are species of *Nereis* (Fig. 43, p. 150) and allied genera and in them the body is divided into a hundred or more segments bearing laterally projecting *parapodia* (Fig. 43d) with projecting bristles. These are the organs of locomotion. They are active animals with an array of tentacles as well as a series of small eyes on the well-developed head. Powerful jaws are protruded at the end of a short muscular proboscis (Fig. 43, b) when feeding on the debris of animal and plant remains which appears to constitute their omni-vorous diet. They range in size from the little purplish-red *Platynereis dumerilii*, some two inches long, to the bronze-coloured *Nereis pelagica* and greenish-grey *Perinereis cultrifera* (Fig. 43, a) more than twice as long, and the impressive green *N. virens* which in some localities reaches lengths of over eighteen inches, and can give the incautious searcher a nasty bite.

These worms spend most of their lives in irregular burrows within muddy sand, rich in organic debris, under stones and boulders. But the needs of reproduction draw many, such as *Nereis pelagica*, annually from obscurity. As the sexual products ripen within the bodies of both males and females, the hinder end of the body, in which these are located, changes in form (Fig. 43, c). The lateral parapodia become larger and their spines bigger and more flattened (Fig. 43, e). The colour of the worm becomes richer, and, especially in the male, the eyes enlarge. This sexual stage is known as the *Heteronereis* (such

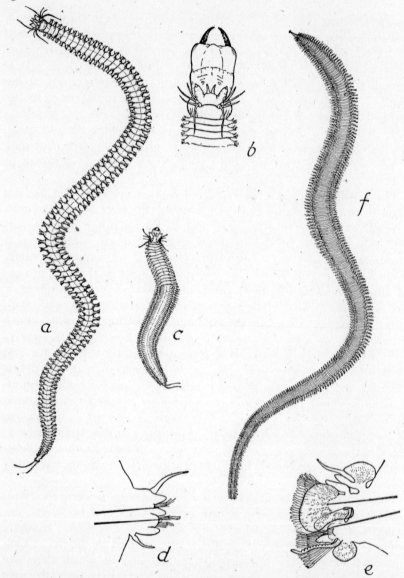

FIG. 43—Polychaete worms. *a. Perinereis cultrifera*, × ¾ ; *b. Nereis longissima*, head and protruded pharynx with jaws, × 6 ; *c. Nereis pelagica*, male in heteronereis stage with posterior half of body modified for swimming, × ¾ ; *d. N. pelagica*, male, normal parapodium, × 8½ ; *e. N. pelagica*, male, parapodium modified for swimming in heteronereis stage, × 8½ ; *f. Glycera* sp., × ¾. (Original drawings by Mrs. E. W. Sexton.)

individuals were once thought to be members of a different genus) and when the change is completed the animals emerge and swim, with the aid of the enlarged, paddle-shaped parapodia, in the water above. There the body wall ruptures to liberate the egg or sperm ; the presence of the sexual products of one sex in the water stimulates the spawning of the other sex so that the sea water soon becomes heavily charged with fertilised eggs which develop into the planktonic larvae. At this stage the creatures possess long spines but there is no segmentation ; this comes later when they metamorphose prior to settlement. Other nereids, *Platynereis dumerilii* for example, incubate the eggs within mucous tubes they construct in the mud. In some this is done by the male after the female has disintegrated following spawning.

In the large paddle-worm, *Phyllodoce lamelligera*, the parapodia are always leaf-like and the animal can swim as well as crawl. But it is usually found under stones near low tide level. It may reach a length of over a foot and has a handsome green or blue, somewhat iridescent, body. The small paddle-worm, *P. maculata*, is commonest on rocks embedded in sand ; it is only three or four inches long and relatively narrower than the larger species.

Less active species include the scale-worms with a protection of flat plates or elytra on the back giving them some resemblance in form, as they may have in habit, to the sea-slaters and the chitons. The commonest species is probably *Polynoë imbricata* which may be found in abundance crawling on rocks. The broad flattened body is an inch or two long. *Halosydna gelatinosa* (Pl. XXa, p. 143), coloured yellow or brown and semi-transparent, is somewhat larger and lives under stones in the *Laminaria* zone. A smaller worm, *Harmothoë lunulata*, hides away in burrows of other animals, sometimes of echinoderms and sometimes of worms such as those of the large red *Amphitrite johnstoni* (Pl. 36b, p. 177) which lives on rocky shores where there is adequate sandy mud for burrowing. The largest of these scale-worms, the sea-mouse, *Aphrodite aculeata*, lives in mud and will be mentioned later.

Of such burrowing worms much the commonest in the localities we are now considering is the redthreads, *Cirratulus cirratus*. When stones are turned to reveal a substratum rich, often odorous, with organic debris, this will often be found covered with very thin, elongated threads of scarlet or orange which are continually twisting. These are the feeding tentacles and gills of the thin-bodied worm, some four inches long, which search will uncover. The tentacles are readily cast off but enough will be retained to reveal their mobile beauty when placed in a bowl of sea water. They collect material from which the edible matter is sorted by flaps guarding the mouth, so that this worm is not an indiscriminate swallower of bottom material like the earth-worm or the lug-worm of sandy shores. *Cirratulus* is faced with the same problem as the ragworms when it comes to spawning. Like them it emerges from beneath the rocks but to crawl upon them and not to swim. Rocks in the Firth of Forth during February have been described as covered with these usually obscurely hidden animals, all ready for spawning. On southern shores the place of *Cirratulus* is usually taken by the allied and larger, though very similar, *Audouinia tentaculata*.

Other burrowing worms which frequently form membranous tubes and may also live in rock crevices include two handsome species, the long and rather fragile red rock-worm, *Marphysa sanguinea*, and the shorter, more brownish *Eunice harassii*, shown in Plates 27b and 28a (opposite, and p. 153). This habit of frequenting rock crevices may well be the origin of the rock-boring habit found in certain small worms. But this higher degree of specialisation will be considered separately in Chapter 12.

There remain the attached worms, namely the serpulids with their limy tubes cemented to the surface of the rocks. The small coiled *Spirorbis* so common on seaweed also occurs on rock (Pl. XXIIIb, p. 158) and with it the much larger *Pomatoceros triqueter* (Pl. XIXa, p. 142), which is easily distinguished by the more or less straight tube with a pronounced keel ending in a sharp spine at the opening. This worm is well worth observing under water when the white tentacles of its crown, spotted with orange, red and blue, are extended together with the opercular plug into which one of them is modified. They need to be observed with caution ; like all such tube-worms, reaction, even to a shadow, is instantaneous withdrawal. There is another species, *Filograna implexa*, which is colonial, with fine interweaving tubes which may form masses on the sides of rocks.

PLATE 27

a. NEMERTINE OR RIBBON-WORM, *Tubulanus annulatus*

b. RED ROCK-WORM, *Marphysa sanguinea*

PLATE 27

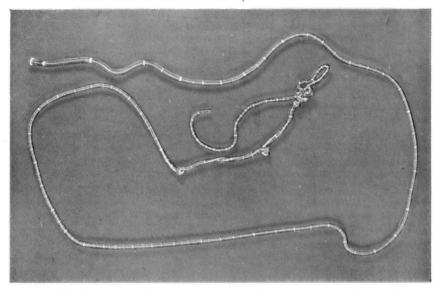

Photographs by D. P. Wilson

PLATE 28

Photographs by D. P. Wilson

Another colonial tube-worm can most suitably be mentioned here although it occurs only in special areas. The honeycomb-worm, *Sabellaria alveolata*, forms massive reefs (Pl. XXb, p. 143) on the exposed surface of rocks where these are bounded near low tide mark by sand. It is not a serpulid, the tube being composed of coarse sand grains which are glued in position by a lobed lip or " building organ," while the operculum consists of two elaborate fans composed of broad, blade-like bristles. The worms cannot leave their tubes but protrude the head and tentacles when the tide is in and so collect both food and the sand grains which are stirred up by the waves. Thus a certain degree of water movement is necessary as well as the presence of sand, on the other hand too great wave action would wash away the colony so that *Sabellaria* reefs are never found in the more exposed areas. The tubes are relatively firm and, although the individual worms are not more than two inches long, they are so numerous and the tubes so closely adherent that masses several feet across and with something of the consistency of porous sandstone are constructed. These are often hollowed out forming small grottoes providing shelter for a variety of other animals. Alike in their powers of building and in the protection thus afforded to other animals, these honeycomb-worms represent the nearest approach on British shores to the reef-building corals of tropical waters.

It is necessary now to turn from worms to crustaceans, of which there is no smaller wealth of species. Stones turned near high water level will disclose a jumping mass of the common shore-skipper, *Orchestia gammarellus*, and lower on the shore less active species of another common genus of amphipods, *Gammarus*. These are usually in pairs with the smaller male held within the bent body of the female which shuffles actively along on its flattened side when disturbed. All are scavengers and feed mainly on decaying weed and other debris. Yet other amphipods live in tubular nests constructed of fragments of weed bound together with threads not unlike the byssus of mussels but produced from glands in certain of the legs. The commonest animal of this habit is *Amphithoë rubricata*. A small isopod, like a

miniature sea-slater but coloured grey, named *Jaera marina*, is also
common under stones.

The best known of the shore crustaceans are the crabs, above all
the common shore-crab, *Carcinus maenas* (Pl. XXIa), which ranges
everywhere from the highest pools down to the *Laminaria* zone and
is largely indifferent to the nature of the bottom and to the degree
of salinity or of exposure. There is no hardier animal on the shore.
It occurs in many colour varieties from blackish-green to reddish and
the broad carapace bears many different patterns. It is a pugnacious
animal and meets all intruders with the opened claws raised and ready
for attack ; *le crab enragé* is its appropriate French name. Nothing
that is edible comes amiss to this most active and ubiquitous of
scavengers. Small individuals will be found inserted under stones,
larger ones in any cranny large enough to harbour them. Near low
tide mark during the summer months small specimens of the edible
crab, *Cancer pagurus* (Pl. 28b, p. 153), are not uncommon and occasionally some large enough for eating. But this is a seasonal visitor which,
like the lobster, in the winter moves offshore where it spawns, the eggs
to hatch out in shallower water during the summer months, whereas
the shore-crab is a constant member of the intertidal fauna.

In certain areas on the west coast, such as the northern shores of
Devon and Cornwall, but also on the shores of Hebridean islands, will
be found *Xantho incisus* (Pl. XXIb), like a small edible crab with a
dark-coloured, rounded shell under two inches wide in the males and
about an inch wider in the females. Spider-crabs of various kinds may
already have been encountered in rock pools, but they occur, though
less commonly, under stones and weed. Finally there are the swimming-
crabs, in which the last two pairs of legs are flattened terminally to form
paddles which permit darting movements through the water. Such
crabs are often thrown in numbers on the shore after storms but
normally live in the sub-littoral zone, though extending into the lower
pools. But the large velvet swimming-crab, *Portunus puber* (Pl. 29,
p. 162), is far from uncommon under weed between tide marks around
our south-western coasts. It is a most handsome crab with a coating
of densely-growing hairs and the nearest equivalent we have to the
large edible swimming-crabs of tropical shores.

These crabs and related decapod crustaceans are always in danger
of having a large claw or one of the walking legs caught under stones,
rolled about by the waves, or seized by some tenacious enemy. Under

PLATE XXI

D. P. Wilson

a. Common Shore Crab, *Carcinus maenas* (× ½)

D. P. Wilson

b. Furrowed Crab, *Xantho incisus* (× ½)

PLATE XXII

D. P. Wilson

a. Spider Crab, *Hyas araneus*. Photographed under water ($\times \frac{2}{3}$)

D. P. Wilson

b. Velvet Swimming Crab, *Portunus puber*, in act of emerging from old shell
at completion of moulting ($\times \frac{1}{2}$)

such conditions they sacrifice the limb so that its owner may survive. This process is known as *autotomy* and occurs in all crabs and in most anomurans, which include the hermit-crabs and their allies, and also in the lobster. It is far from being a simple break at one of the numerous joints in the limb, but takes place along a pre-destined breaking plane, an encircling groove near the base of the third segment of the limb counting from the junction with the body (Fig. 44). There is a special

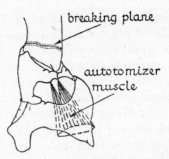

FIG. 44—Basal portion of second walking leg of the lobster, *Homarus*, anterior view, showing mechanism of autotomy, i.e. severance of the limb by the action of the animal. (From Wood and Wood, *Journal Experimental Zoology*, Vol. 62, p. 1.)

autotomiser muscle which comes into play and causes the limb to be bent at such an extreme angle that great pressure is applied along this plane of weakness, which ultimately fractures. The whole process is a reflex action—that is, an automatic response to injury of the limb and it can be produced experimentally by suitable stimulation, or even after death by pulling on the tendon of the autotomiser muscle. It is in no way a deliberate act of the animal but a mechanism outside its control ; an adaptation to ensure survival.

Surprisingly little damage is done to the tissues within the limb ; no muscles are severed and the exposed surface is very small owing to the depth of the surrounding groove. The exposed area is covered with a thin membrane perforated in the centre where the blood-vessel and nerves are withdrawn. This opening is then closed with a clot of blood, for the bluish blood of these crustaceans coagulates on exposure although in a different way from that of the red blood of vertebrates. Over this stump, which gradually turns black, the superficial tissues which form the shell extend and form a small papilla from which a new limb is eventually regenerated. This appears in miniature after the next moult and increases in size with each successive moult until it catches up with its partner on the other side and no evidence remains that it was ever lost. Autotomy is of frequent occurrence and it is not difficult to find crabs and similar crustaceans with partially regenerated limbs.

Under stones where there is mud lives the hairy porcelain crab,

Porcellana platycheles (Fig. 13, p. 31). This is not a true crab, but, as pointed out earlier, is a relative of the asymmetrical hermit-crabs and the squat lobsters, namely an anomuran. It provides an admirable example of adaptation to shore life. The small body and large claws, the latter almost as large as the rounded shell, are flattened and the animal maintains close attachment to the surface of the rock by means of the walking legs which end in sharply pointed spines. Thus secured and offering the minimum of projecting surface, it stands little risk of being washed away even by stormy seas. A smaller species, *P. longicornis*, the minute porcelain crab, lives under stones and among the holdfasts of tangle-weeds, always clear of mud. The shell is only a few millimetres wide and often pale red. These porcelain crabs differ from the true crabs in their mode of feeding. The mouth parts are fringed with hairs which serve to filter fine particles from the water.

Hermit-crabs are so common and well known that we tend to take for granted the extraordinary habit of a crustacean abandoning much of its protective armour to seek adventitious shelter within the empty shell of a univalve mollusc. Agility of movement has been surrendered for the greater protection afforded by the stout molluscan shell. The hermits have the same crawling habit as the squat lobsters, their nearest relatives on the shore, but instead of darting swiftly back when danger threatens, they seek shelter within the shell they inhabit. The last appendages of the soft, asymmetrical abdomen are sickle-shaped and spiny. Mr. R. Elmhirst, who has kept hermit crabs in glass models of snail shells, has made most interesting observations on them. He notes that these terminal appendages serve as a clamp, securing the animal to the central pillar of the shell, while the fourth and fifth pairs of legs act as struts. Movements at some distance are perceived, probably by the eyes, and immediately the animal clamps itself firmly to the shell in preparation for possible withdrawal because the sudden contraction of the abdominal muscles which causes a squat lobster to flick back through the water causes a hermit to enter the protection of the shell. Circulation of water within the shell is maintained mainly by general movements of the body and fouling of this water is prevented by the abdomen being bent forward to the mouth of the shell during defecation.

Ungainly in movement and seemingly overburdened with the weight of the acquired shell, the hermit is yet admirably fitted for life on a rocky shore churned by powerful seas or when it is exposed at low tide. The claws, asymmetrical with that on the right the larger (Fig. 12, p. 30), remain heavily armoured like the carapace and act as an operculum, closing the aperture of the shell when the animal withdraws. Then it can be rolled about by the waves with the same immunity to damage as possessed by the original inhabitant of the shell; and the animal remains moist when exposed to the air. So well can such animals withstand desiccation that well above high water level around tropical shores live the large and handsome species of *Coenobita*, truly terrestrial hermits that may climb into bushes. The story is completed by the still larger " coconut-crab " (*Birgus*) which has abandoned the protection of a foreign shell and redeveloped a limy covering over the abdomen. Nevertheless a limited impression of the success of the hermit-crabs would be left without recording their wide abundance in the sub-littoral zone and down to great depths.

But to return to British shores. The unique habits of these crabs raise problems of growth and of reproduction. Moulting takes place within the shell, first of the abdomen and then of the carapace and legs. Growth appears to be slow but inevitably a stage will come when

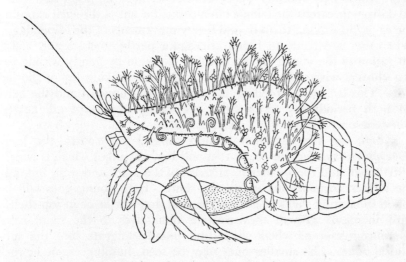

FIG. 45—Encrusting hydroid, *Hydractinia echinata*, spreading over the surface of a whelk-shell inhabited by a hermit-crab. Natural size. (From Allman, *A Monograph of the Gymnoblastic or Tubularian Hydroids*.)

the animal becomes too large for its home. Change must be made to one of larger size and the hermit takes immediate steps to find this. The procedure may well be described by quoting from Gosse's *Aquarium*. After finding a suitable empty shell the hermit-crab " immediately began to turn it about, rolling it over and over with his sharp feet. . . . He carefully examined the interior, feeling it all over with both claws, and trying every spot as far as he could reach ; this examination he continued for perhaps five minutes, and then, as if satisfied, drew out his feet and made an essay to quit his own shell. It was apparent that the exposure of his soft person was considered somewhat dangerous, for he first felt with his antennae in all directions around, vibrating them up and down, and partly coming out and retreating several times before he ventured. At length, however, out he popped, and into the new house as quickly, where he turned and settled himself comfortably."

The developing eggs continue to be carried on the appendages of the abdomen, which indeed have no other function. They are confined to the left side and are correspondingly richly covered with hairs to carry the double burden of eggs. Though well protected, the eggs, which may number many thousands, need aeration and when no danger threatens the female often comes far out of the shell exposing the eggs and waving them to and fro by movements of the appendages. When they are ready to hatch, she again partly emerges and assists liberation of the young larvae into the plankton by gently wiping the egg clusters with a brush of hairs on the last walking leg of the left side. This leg, with its fellow, is normally bent upward and the hairs on both periodically scrub clean the vitally important gill cavity. They are only technically " walking " legs.

There are two common hermits around our coasts, the large *Eupagurus bernhardus* (Pl. 30, p. 163, and Fig. 12, p. 30) which is widely distributed, and the smaller *E. prideauxi* of the south and west, but only the former normally occurs on the shore. Here young ones will be found inhabiting shells of periwinkles and dog-whelks or in top-shells ; with increased growth they transfer themselves to the more commodious shelter of whelk shells and usually migrate into the sublittoral zone. The smaller ones may be seen shuffling about in rock pools or, tucked in their shells, beneath the protective cover of seaweed or stones. In their feeding habits they show an interesting blend of the omnivorous scavenging of the true crabs and the filtering mechanism

PLATE XXIII

D. P. Wilson

a. Sun-star, *Solaster papposus.* Photographed under water ($\times \frac{2}{3}$)

M. A. Wilson

b. Starlet, *Asterina gibbosa,* with Tube-worms, *Spirorbis borealis,* exposed on rock
($\times 1\frac{1}{3}$)

PLATE XXIV

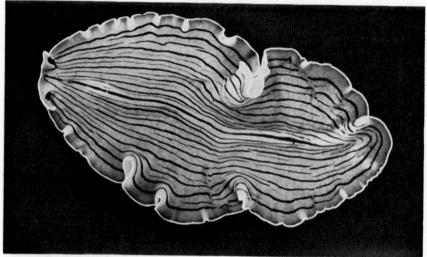

a. Flatworm, *Prostheceraeus vittatus*. Photographed under water (×3)

D. P. Wilson

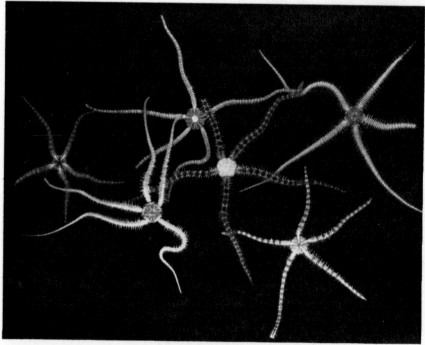

b. Common Brittle-stars, *Ophiothrix fragilis*. Photographed under water (×½)

D. P. Wilson

of the allied porcelain crabs ; they are thus capable of taking food both from the bottom and from the surrounding water.

Hermits seldom live to themselves. The shells, especially the larger ones inhabited by fully grown animals, are usually covered with encrusting barnacles, serpulid worms or saddle-oysters, with polyzoans, hydroids and sponges. Many of these have settled there by chance, as they would on any hard surface, but there are some species which are normally associated with hermit-crabs and seldom, if ever, live elsewhere but on their shells. The shells inhabited by both species may not infrequently be partly covered with a pink moss-like growth of the naked hydroid, *Hydractinia echinata* (Fig. 45, p. 157). This species has the added interest of carrying polyps of various kinds, some for feeding and so bearing tentacles and mouths, others, probably protective, being long and slender and without mouths, and others again being solely concerned with reproduction. All grow from a common basal crust which is also raised into many sharp spines beneath the protective points of which all types of polyps may contract. In the winter the colonies die down to small masses of cells from which new ones arise in the spring. This association can only be of advantage to the hydroid, which has increased chances of obtaining food. Occasionally on *E. bernhardus*, though very commonly on a third invariably sub-littoral species, *E. cuanensis*, there grow masses of the same sponge, *Ficulina ficus*, already reported from the back of the sponge-crab, *Dromia*.

The most conspicuous animals to occur in regular symbiotic association with hermits are anemones. The larger specimens of *E. bernhardus* often carry *Calliactis* (*Adamsia*) *parasitica*, occasionally more than one. This is a typical sea-anemone and is not exclusively found on the shells of hermits ; it may occur on the shell of a living snail or on a crab or even on rocks. When the hermit is feeding the anemone normally bends down so that the tentacles sweep the ground as shown in Plate 30b, (p. 163). There they pick up fragments of food left by the crab. *E. prideauxi* has a constant companion in another anemone, *Adamsia palliata*, the basal disc of which wraps right round the shell, the margins fusing on the far side. It is thus committed to permanent life on that shell. As the anemone grows it increases the effective capacity of the shell so that the hermit never needs to leave it in search of a larger one. This very intimate association of hermit and anemone will rarely if ever be found on the shore but may be seen in the public aquaria which all marine laboratories possess.

A very different animal lives within the shell, one of the ragworms, *Nereis furcata*. It also lives elsewhere and not all, although the majority, of hermit shells harbour this worm. It lies on the right upper side of the crab, in the inward path of the water currents created by it, and may poke out its head when the crab is feeding.

The exact significance of the relationships between the crabs and their various associates has provoked much discussion. It is almost certain that the hermits must gain some added protection from the presence of the anemones with their numerous batteries of stinging cells. The ragworms may be of benefit by keeping the interior of the shell clean and, by their wriggling movements, assisting the circulation of water within it. The anemones would certainly appear to gain by the added opportunities for feeding ; they are carried to many feeding grounds, and there are always fragments to be picked up when the hermit is tearing up its food. The ragworm also may have a better chance of obtaining food while it is certainly protected within the shell. In no case does either party lose in this type of association which is known as symbiosis.

Another type of association, in which one party gains at the expense of the other, is parasitism. This is of wide and perfectly normal occurrence throughout the animal and plant kingdoms. Although the parasite feeds on the host or diverts its food to its own uses and so may affect its growth or, in certain cases, its powers of reproduction, the host animal is not killed and when the parasite dies it resumes growth or reproduction. Hermit-crabs frequently carry parasites in the form of small yellowish bag-like bodies with no apparent structure that hang down on the under-side of the abdomen. This parasite is called *Peltogaster paguri*. A similar parasite, *Sacculina carcini*, is common on shore-crabs, and as it is larger and more easily observed as well as very common it will be described here.

Sacculina forms a rounded yellow mass on the under-side of the abdomen (Fig. 46, p. 161) lying between this and the under-side of the body against which the abdomen is normally closely applied except in females when in berry. The only structure shown by the parasite is a terminal opening, through which the reproductive products are shed. There exists no better example of the effects of parasitic life ; the animal obtains all its food from the host crab and consists only of ramifying rootlets for absorbing nutriment and a protruding mass

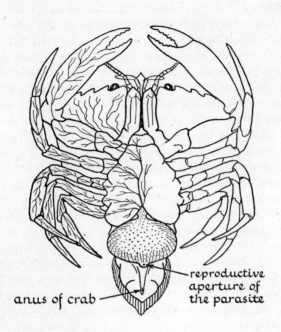

anus of crab

reproductive
aperture of
the parasite

FIG. 46—*Sacculina carcini* parasitic upon the shore-crab, *Carcinus maenas*. About natural size. The crab is viewed from the under-side with the abdomen extended. The parasite consists externally of a rounded mass (shown stippled), containing the reproductive organs, connected with a series of root-like processes which ramify through the body of the crab. Their distribution is indicated on the one side only. (From *Thomson's Outlines of Zoology*, revised by Ritchie, after Delage.)

containing the reproductive organs. Study of such structure as there is tells us nothing of the true nature of such a parasite ; this is revealed by the life-history.

The fertilised eggs give rise in the sea to typical crustacean larvae of a kind almost identical with those of barnacles. Both *Peltogaster* and *Sacculina* are really barnacles in which one-time settlement on the shells of larger crustaceans eventually led to the assumption of a parasitic mode of life with the gradual loss of all structures no longer necessary. The strain on the resources of the host animal is so great as to prevent

moulting and the shell of an infected crab is often covered with encrusting growths, evidence of the length of time since the parasite was
acquired. Reproduction is similarly inhibited, the condition so produced being known as parasitic castration. Eventually, however, the
parasite dies and the crab then resumes growth and reproduction.
Other crustacean parasites found on the shore include isopods which
inhabit the gill chambers of hermits and of prawns, giving the latter a
swollen "face" as though suffering from a gumboil. Common and
horse-mussels often harbour small pea-crabs in the mantle cavity.
These feed partly by intercepting the food of the bivalves and partly
on the soft tissue of the gills. The females grow so large that they are
unable to leave the mussels, but the males are smaller and can move
in and out and so are able to fertilise the imprisoned females.

Little as yet has been said about echinoderms although some
starfishes and sea-urchins may have been seen when examining rock
pools. Starfishes are common between and under rocks although the
species vary with the locality apart from the ubiquitous common
starfish, *Asterias rubens*, which may be purple or orange as well as red
in colour. It is especially abundant on mussel beds where it is shown
in Plate 31 (p. 166). Like all its kind it is carnivorous and feeds mainly
on bivalves. Moving on to one of these, most often a mussel, it humps
the body and grips the two valves of the shell with the tube feet that
cluster along the under-surfaces of the five arms. By a continuous
steady pull it gradually succeeds in overcoming the much greater
initial power of the adductor muscles that pull the valves together.
As a result the shell gapes open. The starfish has no teeth while the
restricted mouth-opening is bounded by the limy plates which form
the skeleton. Unable either to break up or to swallow its prey, it
solves the problem of consuming it by protruding the large stomach
through the mouth and digesting the bivalve externally (Fig. 47, p. 164).

On southern and western shores the larger spiny *Marthasterias
glacialis* (Pl. 32b, p. 167) will be found. This is grey or mauve and
the arms bear rows of prominent spines. *Henricia sanguinolenta* (Pl. 32a),
the so-called scarlet starfish (actually most variable in colour), is
probably commonest on the east coast, but it is a species of wide
distribution and extends to very great depths. It has five arms like

PLATE 29
VELVET SWIMMING-CRAB, *Portunus puber*

PLATE 29

D. P. Wilson

PLATE 30

Photographs by D. P. Wilson

Asterias but is smaller and the arms are thinner. Occasionally on the west coast the large and handsome sun-star, *Solaster papposus* (Pl. XXIIIa, p. 158), which has up to thirteen arms and a broader disc, may appear between tide marks. It may be red or orange but is usually purplish-red with white patches. The allied *S. endeca*, the purple sun-star, less frequently appears on northern shores. It is absent in the south where alone is found our smallest starfish, the starlet, *Asterina gibbosa*. This is seldom much more than an inch across and has a relatively large disc with the five arms so reduced that the animal is almost pentagonal. It is an inconspicuous, dull brown animal and adheres closely to the surface of rocks as shown in Plate XXIIIb.

Brittle-stars are not common on the shore except sometimes as flotsam thrown up after storms soon to become whitened and twisted skeletons but easily identified by the long thin arms sharply demarcated from the disc. On a calm day they may be viewed from a boat through the shallow water of sheltered inlets where they may carpet the sandy bottom. At low water of spring tides the common brittle-star, *Ophiothrix fragilis* (Pl. XXIVb, p. 159), may be found often tightly insinuated under stones or twined among the holdfasts of tangle-weeds. Other species may accompany it, notably in the north, *Ophiopholis aculeata*. Both species vary greatly in colour, the former in particular ranging from a uniform dark violet to red or white with spots or bands of colour on the disc and arms.

Starfishes and brittle-stars are exposed to the same danger of having their arms caught under stones or by enemies as are the larger crustaceans. They counter it by a similar parting with the arm so caught or damaged. This usually takes place near the junction with the disc although there is no invariable region of severance as there is in the crustaceans. Brittle-stars are so named because of the extreme readiness with which they part from their arms ; this occurs even when they are lightly handled so that intact specimens are often difficult to obtain. Starfish are often found with one or more arms shorter than the others but with the stumps, unless the result of very recent mutilation, already sending out a new bud-like growth destined in

---

PLATE 30
a. and *b.* L A R G E H E R M I T - C R A B, *Eupagurus bernhardus*, and " parasitic " Anemone, *Calliactis parasitica*. In the lower photograph the anemone is shown bending down so as to sweep the ground for food with its tentacles

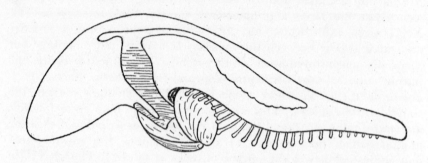

FIG. 47—Diagram showing how a starfish, after pulling open a bivalve by means of its tube-feet, everts its stomach through the small mouth over its prey. (From Russell and Yonge, *The Seas*, after Hirsch.)

time to effect complete regeneration. Even a single arm, if attached to at least half of the disc, may in *Asterias rubens* regenerate the four that are missing and such a " comet " form with one full-sized and four very small arms is occasionally found. Brittle-stars being less common on the shore are less often found with regenerating arms but they have similar powers.

The globular, spiny bodies of sea-urchins are unmistakable. The commonest between tide marks are the small species, *Psammechinus miliaris*, with a maximum horizontal diameter of about two inches, which occurs generally around our coasts, and *Strongylocentrotus dröbachiensis*, which may be half as wide again and is confined to the east coast. Both are green but the former has the characteristic habit of covering itself with pebbles and weed. There is a southern species, *Paracentrotus lividus*, extremely common in the Mediterranean, that occurs on the south and west coasts of Ireland and occasionally around Devon and Cornwall and in the Hebrides. Here the spines are stout and long and capable of making deep cavities in the rock in which the animal lives protected, often in numbers together. It is intermediate in size between the other two and usually brown to violet in colour.

Our large sea-urchin, *Echinus esculentus* (Pl. 33, p. 168), is very common in the sub-littoral region on a hard bottom but of variable occurrence between tide marks. Thus in the Plymouth area it is unknown on the shore but at Millport, on the Clyde, it can be collected in numbers at any good low tide. It has been suggested that the factor concerned is the time of day at which low water of spring tides occurs. Around Plymouth such low tides are about midday, at Millport in the early morning and evening. Hence the shore animals at Plymouth are exposed to extremes of heat at midday and of cold at midnight which are never experienced by the shore fauna at Millport. While animals high on the shore are inured to such extremes, creatures like *Echinus* which are only uncovered at spring tides are not so adapted and this may explain the differences in vertical distribution between the two areas. But there is another suggested explanation which will be discussed in Chapter 18.

Where this urchin does not occur between tide marks the empty tests, often denuded of spines, are not uncommon objects in the flotsam. The complex lantern with its five teeth (Fig. 42, p. 148) will often be found intact within the hollow shell. It is used in browsing on encrusting vegetation and assists the tube feet in slow locomotion. Both tube feet and spines will be seen in an intact specimen. The former, which may extend for lengths of up to two inches, grip so

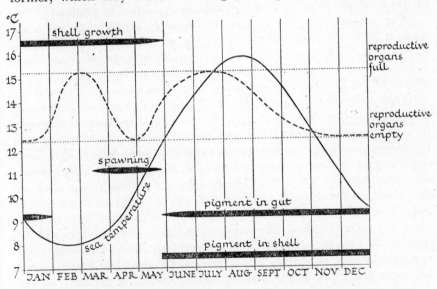

FIG. 48—Generalised diagram showing the seasons of shell and gonad growth, spawning and deposit of pigment in the gut and shell in the common sea-urchin, *Echinus esculentus*, with sea temperature curve (from Plymouth). (Adapted from Moore, *Journal Marine Biological Association*, Vol. XXI, p. 718.)

firmly that it takes a force of from 4 to 6 pounds to dislodge a securely attached urchin. The spines are protective but also assist during occasional movement over a sandy or gravel bottom. The interior contains little beyond the very thin-walled gut and the often voluminous reproductive organs. They are the edible parts of this and other species of sea-urchins which are eagerly sought after as food in the Mediterranean where they have been esteemed since classic times.

Further study of the bared test will reveal that it is made up of very many small plates fitting closely together, but, despite first impressions, this is not an external skeleton like that of a crustacean or a bivalve but an internal one like our own, although covered only by the thinnest film of living tissue. Growth takes place by gradual increase of each plate and careful horizontal grinding of these reveals rings of growth. Like the rings on the scales of fish such as herring or salmon, these permit an estimation of age to be made and it appears that *E. esculentus* can live for eight or more years. But whereas the rings on the scales of fish are due to slower growth in winter, those of the sea-urchin are the result of annual deposition of pigment.

The annual cycle of events in the life of *E. esculentus*, as far as we know it, is shown diagrammatically in Fig. 48 (p. 165) and is somewhat as follows : Shell growth, without accompanying pigmentation, starts in January and continues until about May. Spawning begins in March, when the sea is almost at its coldest, and ends in May. There is no external difference between the sexes and both males and females discharge their products freely into the sea where the larvae, bilaterally instead of radially symmetrical and totally unlike the adults, join the temporary plankton. In the summer months, when food is abundant, the reproductive organs again increase while violet and red pigments accumulate in the gut. From thence they are carried to the shell by cells that wander through the fluid that fills the internal cavity of the test. The shell has now ceased to grow and the white lime of each plate is covered with a thin layer of colour. This in turn will be covered by new depositions of uncoloured lime in the following year, so that each ring represents a thin layer of pigment laid down during the latter half of each year and separating the much thicker layers of

PLATE 31

COMMON STARFISH, *Asterias rubens*, and Mussels, *Mytilus edulis*, with shell valves open under water

PLATE 31

PLATE 32

Photographs by D. P. Wilson

white lime formed in the first half. In January all colouring matter disappears from the gut. Finally the reproductive organs, which swell in the summer largely with food reserves, diminish in the autumn and early winter when these stores are drawn upon, and then enlarge again in February and March in preparation for spawning.

The remaining shore echinoderms will not detain us long. Two species of elongated sea-cucumbers occur although they are inconspicuous. Only one, *Cucumaria lactea*, is widely distributed but it is only about an inch long. *C. saxicola* (Fig. 15, p. 36), which is about three times as long, is restricted to the south-west where it inhabits narrow cracks or holes left by rock borers. Both have tough, flexible skins strengthened with isolated calcareous plates and they are able to grip the rock surface with the short tube feet which are arranged mainly in five rows along the body. The head end is ringed with branching tentacles covered with a sticky substance to which food adheres. From time to time the tentacles are turned into the mouth and " licked " clean as they are withdrawn.

The bulk of the larger molluscs of rocky shores have already been mentioned. There are many small snails, details of which must be sought in books dealing with molluscs ; empty shells of sub-littoral species are common ; nudibranchs such as the sea-lemon and the aeolids already seen in rock pools may also be found under boulders. A few additional animals are worthy of note. One is the common whelk, *Buccinum undatum* (Pl. XXVa, p. 170), which, when fully grown, is the largest of our inshore univalves. But the large shells found on the shore are almost always tenantless ; only small animals live there, and usually where there is mud between the boulders. The shell has the same type of siphonal process as that of the dog-whelk but the actual siphon, or breathing tube, is longer so that the animal can plough through a softish substratum and still draw clean water into the gill chamber. The whelk lays elaborate egg capsules, the female turning slowly round as she produces them so that eventually a large rounded mass of adherent capsules is formed. These masses become detached after the young crawl out and they are common objects

---

PLATE 32
a. "SCARLET" STARFISH, *Henricia sanguinolenta*

b. SPINY STARFISH, *Marthasterias glacialis*

among the flotsam along the strand line. Each is several inches across and like a dried sponge in appearance and texture ; sometimes many are massed together, when they attain the size of an irregular football.

In early summer months a frequent, though somewhat erratic, shore visitor is the graceful sea-hare, *Aplysia punctata*. It has a soft body, olive-green, brown or reddish in colour, which collapses into a shapeless mass out of water but displays the elongated grace shown in Plate 34a (p. 169) when returned to it. Two pairs of tentacles are carried on the head, of which the upper pair, in their resemblance to the ears of a hare, are responsible for the common name. Despite the soft body, *Aplysia* is not a true sea-slug because it has a shell, although this is thin and transparent and almost entirely enclosed within the tissues. Moreover it has a gill cavity and a true gill. It is an example of a tectibranch snail of which we have various representatives in our fauna. They form a link between the shelled snails and the true sea-slugs. The sea-hare comes inshore to spread tangled threads of salmon-coloured spawn (shown in Pl. 34a, p. 169) on the surface of rock and weed. In feeding habits it is in startling contrast to the whelk. The latter is a scavenging carnivore which cleans out the bodies of dead or moribund animals with its long proboscis and may also attack living bivalves. The sea-hare is exclusively herbivorous and confines itself to the sea-lettuce, *Ulva*, biting off sizable pieces with its large jaws, passing these by the conveyor belt of the lingual ribbon into the gut and there reducing them to fine fragments by trituration in an internal gizzard. Life is short ; the young hatch out from the strings of spawn, settle from the plankton some distance offshore and move inshore as they attain adult size and finally, at the end of a year, pass on to the shore for spawning, after which they die.

Reduction and overgrowth of the shell is not confined to the tectibranch snails. It occurs also in *Lamellaria perspicua* (Pl. 38b, p. 245) which is in other respects more closely allied to the periwinkles. This animal is not uncommon under stones on the shore and feeds on the encrusting growths of compound sea-squirts which it rather resembles in its variable colour, most usually a mottled yellow. Shore specimens seldom exceed an inch in length and are often smaller ; they are most,

---

PLATE 33
EDIBLE SEA-URCHINS, *Echinus esculentus*, viewed from the side and above and showing the spines and the extended tube feet

PLATE 33

PLATE 34

Photographs by D. P. Wilson

easily distinguished from true sea-slugs by the presence of a small siphonal process in the mid-line just behind the head.

The common shore bivalves are usually attached by byssus threads like the common mussel. Great exposure at low spring tides sometimes reveals the characteristically sub-littoral scallops which are not attached except when very small and do not burrow like the majority of bivalves. Scallop shells are too well known to demand description ; empty valves are commonly washed up on the shore while both the large *Pecten maximus* (Pl. 34b) and the smaller queen-scallop, *Chlamys (Aequipecten) opercularis*, are highly prized as food and frequently offered for sale. No bivalve shells have greater beauty of design, a design which has been copied by man since the dawn of history. The scallop-shell, in this case *Pecten jacobaeus* of the Mediterranean, was the badge of the pilgrim. The interested reader will find a well-documented account of the possible origin of this custom and of its later developments in *The Edible Mollusca of Great Britain and Ireland* by M. S. Lovell.

Scallops have many unique features. They lie with the shell valves horizontal instead of vertical ; in *P. maximus* the right valve, on which the animal lies, is rounded and the left one flat. The animal is thus raised above the bottom and draws in correspondingly clearer water. The original two adductor muscles of the bivalves, already in the mussels an asymmetrical pair, have become further reduced in the scallops to a single large muscle which occupies the centre of the rounded shell. The hinge-line is long and straight and bears no teeth but a stout elastic ligament which causes the valves to gape widely apart when the mussel is relaxed. When the shell is open there can be seen just within the margin of each valve a line of small tentacles and between these many fine glistening points. Each is an eye with well-developed focusing lens, receptive retina and conducting nerve fibres. A little further within is a deep curtain, dropping down from the upper valve and projecting up from the lower one. These jointly control the inflow and outflow of water.

PLATE 34 .

a. SEA-HARES, *Aplysia punctata*, with coil of their pink spawn in foreground, and behind the seaweeds, *Ulva* (green), *Fucus serratus* (brown), *Polysiphonia* (red)

b. SCALLOP, *Pecten maximus*, with seaweeds, *Laminaria* and *Ulva*, attached to the upper, flat shell valve

So equipped the scallops have acquired what is for a bivalve the surprising capacity of swimming. Being free, any sudden ejection of water from the gill cavity when the central muscle contracts causes the animal to move in the opposite direction. The position where water is expelled depends on the disposition of the curtains of tissue which can be extended or withdrawn locally. The different types of movement of which the smaller queen-scallop is capable (and it is a much more active animal than the larger species) are shown in Fig. 49 (p. 172). In the typical swimming movement, the animal flaps the shell valves vigorously, leaving the bottom and moving upward in a series of convulsive jerks corresponding to the closures of the shell and falling down a little between each movement. The free edge of the shell goes in front so that the animal appears to be taking a series of bites out of the water. The flat, streamlined shell is almost horizontal so that it sinks but little between each movement, with the hinge region a little depressed. As shown in Fig. 49, a and d, water is expelled backward at each side of the hinge-line and also to a minor extent downward around the free margins of the shell, owing to the overlapping of the lower by the upper curtain. The former provides the force for horizontal movement, the latter both counteracts the effect of gravity and raises the animal in the water.

By such means the scallops move about freely. We know that they migrate because they are not found in the same regions throughout the year, but we are uncertain exactly how far they can move. They can also execute other movements. When lying quietly on the bottom they may be stimulated into sudden activity by the approach of an enemy. They then carry out what is known as the escape movement. The muscle contracts suddenly but the curtains around the free margins of the shell are drawn in so that water is expelled between them and the animal shoots back with the hinge foremost (Fig. 49, b and e). There is no downward ejection of water so that the scallop does not rise from the bottom. The animals may also spin round on the central axis by expelling water from one side only of the hinge line (Fig. 49, c). Finally the larger species in particular can turn itself over if wave action should tumble it over on to the flat valve. This it does by ejecting water downward by appropriate overlapping of the curtains around the free margin, the animal then turning a somersault over the hinge-line (Fig. 49, f).

It is impossible to dismiss the bivalves without mentioning the

PLATE XXV

D. P. Wilson

a. Common Whelk, *Buccinum undatum*, with siphon, head and foot, bearing operculum on upper surface, all protruded under water (×⅔)

D. P. Wilson

b. Gooseberry Sea-squirt, *Dendrodoa grossularia*. Photographed under water (×1)

PLATE XXVI

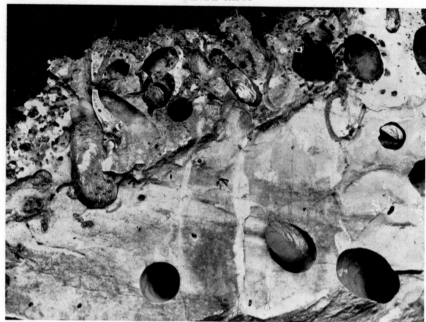

D. P. Wilson

a. Limestone split open to show deep burrows of Bivalve, *Hiatella arctica*; U-shaped
burrows of the Polychaete Worm, *Potamilla torelli*; and superficial honeycombing of the
Boring Sponge, *Cliona celata*. Photographed under water (× ⅓)

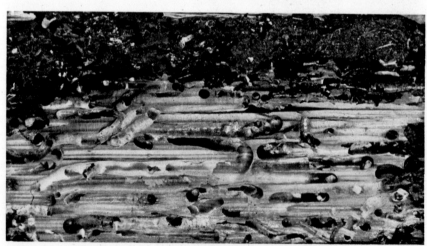

D. P. Wilson

b. Portion of a jetty pile split open to show burrows of the Shipworm, *Teredo navalis*
(× ½)

file-shell, the most beautiful of British species although of somewhat rare occurrence on the shore. Fortunately its beauty has been well displayed in photographs by Mr. D. P. Wilson reproduced in Plate 35 (p. 176). *Lima hians* has much in common with the scallops, including some capacity for swimming although with the valves disposed vertically. But it is normally of very sedentary habits. It is especially common among the holdfasts of tangles and in holes among rocks in the sub-littoral zone but with good fortune may be found exposed at low water of spring tides. It has the unique power among bivalves of constructing an elaborate nest with byssus threads. Holes are left for the entrance and exit of water and the animal lies in reasonable safety within, collecting plankton from the water stream. The need for such protection is shown in the photographs. The tissues are fringed with very many long trailing tentacles of a beautiful red or orange colour. These cannot be withdrawn within the delicate ribbed white shell and the animal is easy prey for the first predator that encounters it outside the protection of the nest. Within the nest the tentacles are so disposed as to guide the water currents. A swimming *Lima* is a most charming sight. Progress is made with a series of rather languid movements with each of which the long fringe of tentacles slowly rises and then gently descends around the white shell.

Apart from the compound forms, the commonest and most conspicuous of the sea-squirts is *Dendrodoa (Styelopsis) grossularia* (Pl. XXVb, p. 170) which resembles a red gooseberry in both colour and shape although the two small projections from the leathery test reveal its true character. It is a simple sea-squirt but generally grows in large clusters which are widely distributed on the shore usually on the sides of stones and sometimes half-covered with mud. Larger sea-squirts include the pale-green, semi-transparent *Ciona intestinalis* which grows to lengths of several inches in sheltered clefts or protected under boulders, and the tougher, opaque *Ascidiella aspersa* (Pl. 25b, p. 144) which is common on the south coast. Among colonial forms the commonest are species of *Morchellium* and *Amaroucium*, not easily distinguished from one another : both are shown in Plate 38b (p. 245). The colonies are sometimes flat and sometimes bun-shaped and vary considerably in colour. Finally there are the beautiful stalked individuals of *Clavelina lepadiformis* (Fig. 19, p. 40), somewhat under an inch high, of delicate texture and orange colour, which die down in the winter to simple resting masses of unorganised cells, from which

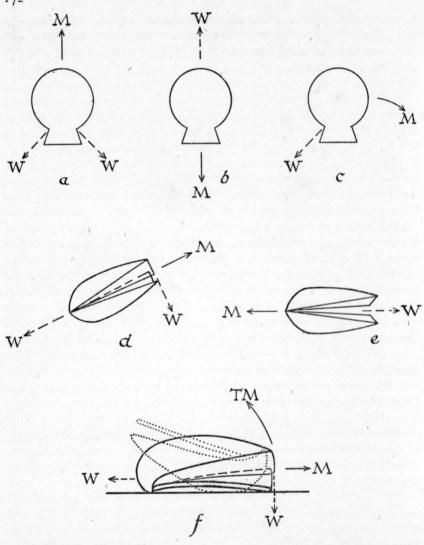

FIG. 49—Movements of scallops : *a-e.* the queen-scallop, *Chlamys opercularis* ; *f.* the large scallop, *Pecten maximus.* Direction of movement (M) indicated by complete arrows and direction in which water expelled (W) by broken arrows. *a.* swimming movement viewed from above ; *b.* escape movement viewed from above ; *c.* twisting movement ; *d.* swimming movement viewed from the side showing disposition of the mantle curtains ; *e.* escape movement viewed from the side ; *f.* mechanism of turning movement (TM) viewed from the side, final position of the scallop being shown by the dotted outline. (Modified after Buddenbrock.)

new individuals arise the following year. Among fish there are the butter-fish and the other blennies which, as we have already seen, may commonly be found when the tide is out, but well protected under stones or weed.

CHAPTER 12

BORERS INTO ROCK AND WOOD

"See ! here is the soft red sandstone lying in great beds, pierced through and through with smooth round holes, just as if bored with a carpenter's auger, big enough to admit a man's thumb. What agency has been in operation to effect these perforations? Let us try to discover."

P. H. GOSSE : *A Year at the Shore.* 1865

MENTION has been made of one animal, the sea-urchin, *Paracentrotus*, which excavates rounded burrows in rock. This it does, where there is need for protection against the force of the sea, only between tide marks and where there is no natural shelter in crannies or deep pools. The urchin usually first occupies a shallow depression and enlarges this by the combined action of teeth and spines, the former deepening and the latter widening the opening. Should the burrow be deep the animal, by subsequent growth, may eventually be trapped within it but this is not usual. *Paracentrotus lividus* is a southern species and only extends on to the more outlying western shores of the British Isles. There are other more widely distributed and more highly specialised rock borers which remain to be described and with them other animals that burrow into wooden pier piles and boats and into floating timber, in which they may be found after this has been cast up on the shore.

A surprising diversity of animals bore into rock. They include sponges, annelid worms, barnacles, isopods and bivalve molluscs, in addition to the sea-urchins. Boring barnacles are confined to tropical shores where they penetrate deeply into dead coral rock, while certain isopod crustaceans burrow into soft rocks of claystone and limestone in New Zealand and Australia. It is surprising to find an immobile animal like a sponge making its way into rock but this is one, although not the invariable, habit of the boring sponge, *Cliona celata*. On sandstone this grows as rounded yellow masses but it bores into limestone (Pl. XXVIa, p. 171) and into the shells of molluscs and may so become a serious pest on oyster beds. The method of boring remains something of a mystery but probably the limestone is dissolved by

PLATE XXVII

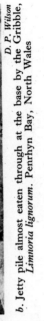

D. P. Wilson

b. Jetty pile almost eaten through at the base by the Gribble, *Limnoria lignorum*. Penrhyn Bay, North Wales

D. P. Wilson

a. Wood from jetty pile split open to show early infection with *Teredo navalis*. The anterior half of one specimen is exposed with the under side of the shell valves and the rounded sucker-like foot at the end of the burrow (×1)

PLATE XXVIII

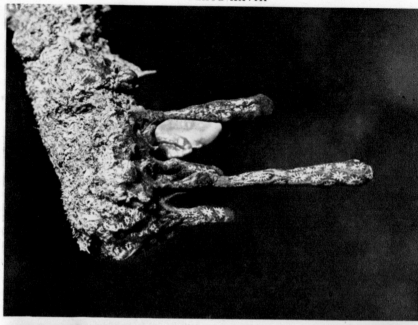

b. Compound Sea-squirt, *Botryllus schlosseri*, growing in sheltered water in Millbay Docks, Plymouth (×3)

D. P. Wilson

a. Colony of Dead Man's Fingers, *Alcyonium digitatum*, showing the expanded polyps. Photographed under water (×¾)

D. P. Wilson

acid. The burrows are never deep, but, as shown in Plate XXVIa, the superficial layer penetrated by them becomes honeycombed to such an extent that a badly attacked oyster shell, covered with a thin yellow crust of sponge, may be literally crumbled between finger and thumb.

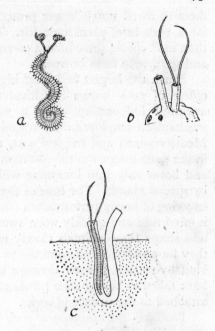

The rock-boring worms are small but common, especially species of *Potamilla* (Pl. XXVIa) and of *Polydora*. The latter construct little projecting tubes of mud (Fig. 50), through one of which two tentacles project. Search will often reveal these worms on the floor or sides of shallow pools in limestone rock though they are also reported from shales and sandstones. They penetrate between the shell valves of oysters and there accumulate mud between the tissues and the shell. Over these the oyster lays down thin layers of shell, forming blisters, which may eventually prevent the valves from closing and so cause

FIG. 50—Rock-boring polychaete worm, *Polydora ciliata* ; *a*. worm extracted from tube ; *b*. mud tubes erected at two openings of burrow, two tentacles extending from one ; *c*. diagrammatic section of burrow with worm in position. Variously magnified. (From Calman, *Marine Boring Animals*, British Museum (Nat. Hist.) Economic Series, No. 10.)

death. Boring is probably assisted mechanically by the stiff bristles but again acid would seem to play the major part in the penetration of limestone. There is an interesting difference between the life-histories of the two common species. In both the developing eggs are first attached to the wall of the burrow, where they are aerated by a current of water created by the movements of the parent worm. The larvae of one species, *P. ciliata*, emerge at a relatively advanced stage but still have to spend something like a month exposed to dangers of life in the plankton before they are able to settle and begin boring. In the other species, *P. hoplura*, the young are provided with additional food in the form of yolk-masses which they swallow and which enable them to dwell much longer protected in the burrow before emerging for a very brief planktonic life. Though fewer young are produced than in *P. ciliata*, these have a correspondingly better chance of survival and settling to form burrows.

Much the largest and most highly adapted and so the most interesting of rock borers are bivalve molluscs. This habit has been acquired independently by a number of these animals and the mechanisms employed vary considerably. In the date-mussels of the Mediterranean and tropical seas, the animal is attached by a byssus, in the same manner as the common mussel, to which it is closely related, and bores only into limestone which it dissolves with acid produced by special glands. The lime of the shell is protected by a thick brown covering of horny periostracum (the outermost layer of the shell which is often thin and quickly worn away in other bivalves). In size, colour and shape these animals closely resemble dates. Within the burrow they lie protected and continue to feed in the normal manner of their kind by straining water through the gills. They may be assumed to have taken to boring after previous existence, like the common mussel, attached to the surface of rocks.

Probably the commonest rock borer around our coasts is the red-nose, *Hiatella* (*Saxicava*) *gallicana*, which, possibly with the smaller *H. arctica* (Pl. XXVIa, p. 171), burrows in limestone and occasionally in sandstone. It also lives on the surface attached by byssus threads to rocks, or among mussels and within the holdfasts of tangle-weeds and is therefore not so specialised a borer as the date-mussels or as the piddocks, which will be described later. The shell is strong and usually very irregular but shows no special modification for boring. This appears to start when the animal becomes attached in a crack of the rock and then to proceed by purely mechanical means. There is no evidence of any acid. The precise mode of boring is not too obvious. Certainly the byssus is not employed for holding on to rock during boring as it is in the date-mussel nor is the foot as it is in the piddocks. The animal appears to obtain the necessary power for boring by taking in water through the muscular siphon which extends from between the hind ends of the shell valves, then closing the openings

PLATE 35

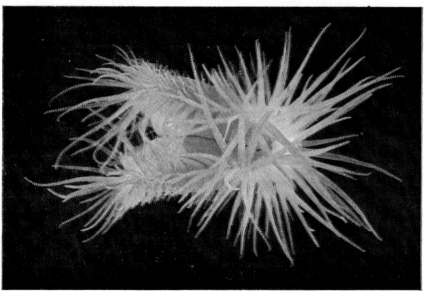

Photographs by D. P. Wilson

PLATE 36

D. P. Wilson

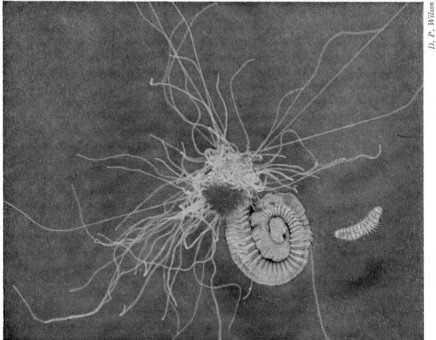

D. P. Wilson

and contracting the walls of the siphonal tubes so that water is forced into the gill cavity between the shell valves. These are pushed apart and the rock surface is abraded by them. Despite this apparently crude method of attack, deep and accurately cut burrows are excavated in hard limestone as shown in Plate XXVIa.

Much the most specialised and efficient rock-borers are the piddocks. They always burrow, and to quote Gwyn Jeffreys, "in the holes thus excavated they dwell at ease, never of their own accord moving from one place to another." They include the large *Pholas dactylus* with a shell up to six inches long and, owing to the great length of its massive siphons, capable of constructing a burrow a foot long, and also smaller species of *Pholadidea*, *Barnea* and *Zirfaea*. *Gastrochaena dubia*, which occurs on the south and west, has similar habits. All bore mechanically and are thus not conditioned by the composition of the rock, although they usually inhabit softer rocks such as sandstones, shales, slate, chalk or even peat and clay. The openings of the burrow to which the tips of the siphons extend are relatively small because the animal enlarges the cavity within as it grows. When these openings are detected exposure of the enclosed animal by means of the Victorian hammer and chisel is well justified.

The shell (Fig. 51, p. 178) is surprisingly thin and brittle but extremely hard and bears rows of fine teeth used in boring. It differs in important respects, all bearing on its acquired function as a boring tool, from that of a typical bivalve. There is no ligament binding the shell valves while the hinge teeth are reduced to a rounded ball on each valve, the two acting like a double ball joint. From the hinge region there descends a short blade-like projection the function of which will become apparent later. In some species, including the large *P. dactylus*, additional shell plates are inserted on the upper side between the two valves (Fig. 51). On the under-side the valves are widely separated in front and expose the rounded and sucker-like foot. This very plastic organ, which is used for locomotion in sand- and mud-

---

PLATE 36
a. L U G - W O R M, *Arenicola marina*

b. T E R E B E L L I D W O R M, *Amphitrite johnstoni*, dug out of mud together with the S M A L L S C A L E - W O R M, *Harmothoë lunulata*, which lives commensally in the burrow of the larger worm

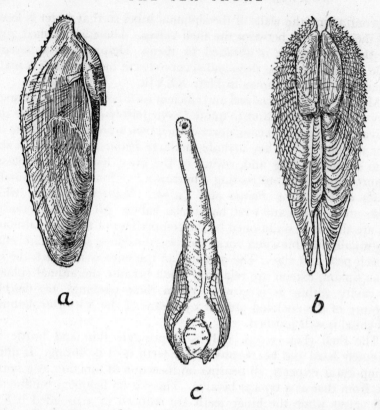

a

b

c

FIG. 51—Rock-boring bivalves : Piddock, *Pholas dactylus, a.* interior of right shell valve ; *b.* complete shell viewed from above ; *c. Barnea parva,* complete animal viewed from below showing extended siphons and rounded, sucker-like foot. Natural size. (From Jeffreys, *British Conchology.*)

living bivalves and also for planting the byssus threads in the mussels, is here modified as an organ of attachment and enables the piddock to take a firm grip of the end of the burrow. The necessary additional attachment for the powerful muscles needed to maintain this suction is provided by the projecting blades of the hinge region.

Boring is effected by alternate contractions of the two adductor muscles running between the front and hind halves of the shell valves. The absence of a ligament and the presence of the ball joint permits a sea-saw movement of the valves. The normal opening and closing action of the typical bivalve is not needed when the animal is already well protected in a deep rock burrow. The teeth on the shell are largely restricted to the front half of the valves so that when these are pulled apart by contraction of the hinder adductors the teeth rasp the surface of the rock and boring proceeds. During this laborious process the animal continually alters the position of the foot, turning first to one side and then to the other, so that an even round-bore hole is cut with slow efficiency through the rock. A strange, and somewhat unaccountable, feature about this species is its phosphorescent properties. Examined in darkness the outlines of the body glow with a greenish-blue light due to the production from special glands of a luminous slime.

Most of the piddocks appear to continue boring throughout life, with the exception of *Pholadidea loscombiana* (Fig. 52). This species occurs near low tide level along the south-west of England and around Ireland and is easily recognised by the trumpet-shaped distension at the end of the shell in which the base of the siphons is protected. After attaining a length of about an inch and a half the animal stops boring. The foot with its powerful muscles then atrophies and the opening through which it formerly extended is reduced to a fine aperture by ingrowth of the surrounding tissues, which are covered externally by a horny layer. The object of boring has been achieved by the attainment of complete protection within the rock and the animal now lies passive in the burrow, content to feed and reproduce itself.

Mention should be made of a small bivalve, *Petricola pholadiformis*, very like a piddock in general appearance and with the same habits. This is an American species which was probably introduced with imported oysters relaid on beds around

FIG. 52—Rock-boring bivalve, *Pholadidea loscombiana*, viewed from the side, showing extended siphons and trumpet-shaped distension at the end of the shell into which they can be withdrawn. Natural size. (From Jeffreys, *British Conchology*.)

the east and south-east coasts of England. It was first noted there round about 1890. Despite its superficial resemblance to the piddocks, this animal belongs to a very different group of bivalves; but little appears to be known about the method in which it bores.

Wood-boring animals consist of bivalves and crustaceans. As the former are allied to the piddocks they may suitably be described first. Most notorious of wood-borers is the shipworm, "calamitas navium" of Linnaeus, known and dreaded since classical times when it riddled the planking of Greek triremes and Roman galleys. Later it was to destroy Drake's *Golden Hind* and, in 1730, to threaten the very existence of Holland by attacking the dykes. It still remains a menace to unprotected boats and to pier piles; between 1914 and 1920 its sudden spread throughout San Francisco Bay caused damage estimated at some ten million dollars.

There are many species of shipworms in different seas, of which three occur in British waters, *Teredo navalis* (Pl. XXVIIa, p. 174), *T. norvegica* and *T. megotara*. The last of these is found only in floating timber together with the similar but less highly modified borer, *Xylophaga dorsalis*, which will be described later. No bivalves are more specialised in structure and habit than shipworms and it is not surprising that they were not even identified as molluscs until 1733 when they were first carefully studied by the Dutch zoologist, G. Snellius, appointed to inquire into the damage done to the dykes. The body is long and worm-like and the small shell confined to the tip (Fig. 53). The structure of the internal organs and of the shell is essentially the same as in the piddocks but represents the result of further evolutionary change.

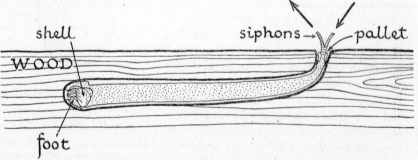

FIG. 53—Shipworm, *Teredo navalis*, diagrammatic section of burrow with animal within it, arrows indicating the direction of water-flow through the siphons.

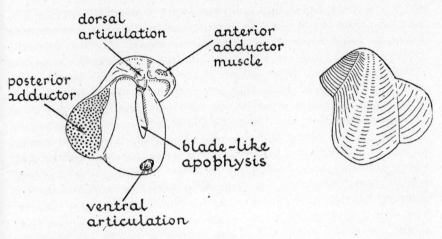

FIG. 54—Left shell valve of shipworm, *Teredo navalis*, internal view on left, external view on right. About twice natural size. (After Miller.)

Teredo starts life as a larva in the plankton. It is equipped with a minute bivalve shell and is only to be distinguished by an expert eye from the larvae of other bivalves. Should water movements carry it by chance against a wooden surface, there it remains. It can be shown experimentally that larvae which, by chance movements, enter fine tubes containing an extract of wood remain within these, whereas they as readily leave as enter similar tubes containing sea water carrying in solution a variety of other, equally innocuous, substances. Once in contact with wood the larvae begin to change into the adult form. A small foot appears and a single byssus thread with which temporary attachment is maintained. The shell now alters in form and in function, ceasing to be a protective covering as it is converted into a most efficient cutting tool. As shown in Fig. 54, the valves can be divided into three regions, a hinder, wing-like lobe known as the auricle, a middle semicircular area which forms the bulk of the valve, and a triangular anterior lobe which extends for only about half the breadth of the middle lobe. The outer surfaces of the middle and anterior lobes are covered with rows of sharp ridges which are continually being added to during the active life of the animal. As in the piddocks, there is no ligament and the hinge teeth are converted into similar rounded surfaces on which the valves rock, but in the shipworms there is a second ball joint on the under-side of the middle lobe, i.e. in the region where in normal bivalves the two valves separate under the influence of the elastic ligament above. The two adductor muscles differ greatly in size; the hinder one is very large and is attached to the extensive surface of the auricles while the anterior one is reduced. The blade-like projection from the hinge region is even better developed than in the piddocks and serves the same function of providing attachment for the muscles operating the foot.

This organ is round and grips the head of the burrow by suction, further grip being provided by a flap of tissue that overlaps the shell valves above as shown in Fig. 53 (p. 180). The shell is thus pressed tightly against the wood. The powerful posterior adductor muscle now contracts; it thus draws the auricles together and forces the anterior lobes apart so that the ridges of sharp teeth on their outer surfaces scrape the wood at the end of the burrow while the ridges on the middle lobes widen the opening behind. The small anterior adductor then contracts in its turn and draws the corresponding lobes together. No rasping of the wood is produced by this movement, which explains the much smaller size of this muscle. After each movement of the valves the foot loosens its hold and moves a little distance round in one direction. This alternation of scraping action and lateral movement continues until the front end of the animal has moved through 180 degrees when the foot reverses the direction of movement. A smooth and perfectly circular burrow is thus cut through the wood.

As the animal bores so does the body elongate; this must be the case because, unlike the piddocks, the shipworms are attached to the wall of the burrow at the hind end of the body. Two new structures are found there in the form of a pair of paddle-like limy plates called the pallets (Fig. 53). These are withdrawn when the siphons are extended through the minute openings, each lined with a thin layer of lime, which are the only external evidence of the presence of animals up to a foot long within the wood. When the siphons are withdrawn the pallets are pushed forward so that their ends meet and effectively close the openings. Water may thus be retained within the burrow for long periods so that the animals are able to survive even after the wood has been out of water for several weeks. Attachment to the walls of the burrow is essential for the action of the pallets by providing the necessary purchase for the muscles which operate them. The remarkably high degree of specialisation achieved by the shipworms has involved complete dependence on life in the original burrow. Unlike piddocks they cannot, if removed, form a new one.

The greatly elongated body harbours the equally drawn-out, and much reduced, gills which create the usual water currents. But the animal makes less demand on the sea water for food than do the rock-borers. Shipworms obtain food as well as protection from the wood. All fragments scraped off by the shell valves in boring are passed into the mouth and through the length of the gut before reaching the exterior by way of the anus and the outgoing siphon. During passage through the gut some at least of the cellulose in the wood is digested by the animal and converted into the soluble sugar, glucose, which provides the chief source of energy in all animals. This power of breaking down the very complex woody constituents of plants is rare in the animal kingdom and largely confined to certain snails and some wood-boring insects. It represents a high degree of functional adaptation and enables the shipworms to obtain the energy needed for boring from the very substance into which they bore.

Shipworms have only a short life, probably not more than one year, but in that time they may make extensive burrows. In newly infected wood these run along the grain, the animal taking the line of least resistance, but where infection is intense the mass of wood is everywhere riddled with burrows which twist and turn in all directions (Pl. XXVIb, p. 171). No two burrows ever meet : the animals are apparently able to detect the proximity of another burrow and turn to avoid it. If unable to continue burrowing in any direction they proceed to lay down a thin layer of lime over the internal surface of the burrow and in this shelly case may continue to live and reproduce for some time, in much the same manner as the adult *Pholadidea*. The piddocks, it may be noted, do not hesitate to invade the privacy of another burrow. Gosse, in his *Year on the Shore*, quotes the statement of a Mrs. Merrifield to the effect that " A lady, watching the operations of some Pholades which were at work in a basin of sea-water, perceived that two of them were boring at such an angle that their tunnels would meet. Curious to ascertain what they would do in this case, she continued her observations, and found that *the larger and stronger Pholas bored straight through the weaker one*, as if it had been merely a piece of chalk rock."

Floating timber is sometimes found bored with animals bearing the same type of shell as *Teredo* but without the long body, merely with siphons that can be withdrawn within the protection of the shell. This animal, *Xylophaga dorsalis*, represents with unusual exactitude a stage through which the shipworms must have passed in their evolutionary history. The burrows it makes are spherical and shallow, no deeper than the diameter of the shell, which is about half an inch. The animals do not digest wood and resemble the rock-borers in obtaining protection in their burrows but food from the sea. Like the shipworms, incidentally, they change sex from male to female as they grow, but, by what appears to be a unique arrangement, they retain the sperms produced during the male phase to fertilise the eggs produced in the later female phase.

The crustacean wood-borers are extremely common and are to be found on most pier piles that have stood for any length of time in the sea (Pl. XXVIIb, p. 174). There are two of them, an isopod and an amphipod, of which the former is much the more important and almost as destructive as the shipworms. This is the gribble, *Limnoria lignorum* (Fig. 55), only about one-sixth of an inch long but usually present in enormous numbers. It has seven pairs of short legs ending in sharp claws with which it grips the sides of the burrow. It bores with the aid of the mandibles which border the mouth. These are asymmetrical, as shown in Fig. 56 (p. 185), that on the right side having a sharp point and a roughened edge which fits into a groove with a rasp-like surface on the left mandible. The two work together like a combined rasp and file.

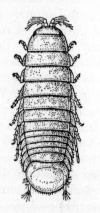

FIG. 55—Gribble, *Limnoria lignorum*. About ten times natural size. (From Calman, *Marine Boring Animals*, after Sars.)

The gribble makes superficial burrows. It hollows out a depression in the soft part of the grain and then passes by an easy incline into the wood and continues to tunnel under this parallel to the surface at a depth of about one-fifth of an inch and for a length of up to one inch (Fig. 57, p. 186). The course of each burrow can be followed during early infection. when the surface of the wood is still relatively intact, by a series of fine holes, like miniature "man-holes," which the animal cuts through the roof as it proceeds and through which a current of water for respiration is drawn. There are normally a pair of gribbles in each

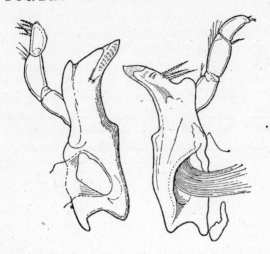

FIG. 56—Mandibles of the gribble, *Limnoria lignorum*, showing " rasp-and-file " combination. Greatly enlarged. (From Russell and Yonge, *The Seas*, after Hoek.)

burrow ; the female is normally at the blind end and appears to do all the work. If disturbed, the male tends to crawl slowly backward out of the burrow but the female grips the sides tenaciously and resists attempts to pull her out. The females are probably in the habit of gripping tightly to provide the necessary purchase when burrowing.

The shipworms, as we have seen, infect new wood by way of the freely swimming larvae. The gribble, like all isopods, carries the large eggs in a brood pouch under the body and from this some twenty to thirty young animals, about one-fifth the size of the adults, eventually emerge. Gravid females and young are found at all seasons but especially in spring. The young immediately begin boring on their own account by hollowing out little burrows from the sides of that of the parent. Hence infection spreads rapidly and when the outermost layers of wood break away the animals burrow into the lower layers until eventually a wide and deep area is eaten away. The best one can say of the matter is that the activities of the gribble are obvious, unlike those of the shipworm which may continue undetected until the whole of the timber is rotten and suddenly collapses. *Limnoria* is always most abundant about low water level so that pier piles are eaten away most rapidly at this level where they are tapered and may

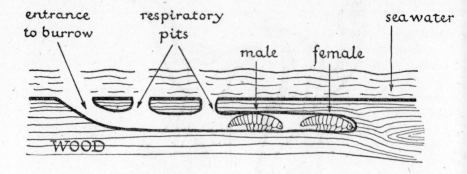

FIG. 57—Gribble, *Limnoria lignorum*, diagrammatic section of burrow with animals in position. (Modified from Russell and Yonge, *The Seas*.)

finally break through, as well shown in Plate XXVIIb (p. 174). Up to four hundred animals have been collected from a square inch of timber at the low water mark.

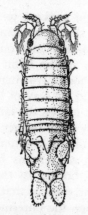

FIG. 58—Wood - boring amphipod, *Chelura tere-brans*, female ; the male differs in having much longer tail-like appendages and dorsal spine. About eight times natural size. (From Calman, *Marine Boring Animals*, after Sars.)

Infection of new wood is always by way of the adults which can swim or crawl from one piece of timber to another. This occurs almost entirely during the winter and the spring. There is no evidence that wood itself is digested although the gribbles may obtain some of their food from the microscopic algae and the bacteria which multiply on rotting wood. They bore primarily for protection and have been found in such a non-nutritious substance as the insulating covering of submarine cables.

Chelura terebrans (Fig. 58) is the amphipod wood-borer. It is somewhat larger than the gribble and has the typical lateral flattening of its kind. It occurs normally with *Limnoria* and would appear to rely to a large extent on the pioneer activities of that animal because it always lives in the more superficial layers of the wood which have already been honeycombed with the formation of channels and pits in which it can live protected. It probably enlarges pre-existing cavities but it is doubtful whether this animal can excavate a burrow unaided by the previous activities of the gribble.

Many methods have been devised for protecting timber from the ravages of these wood-borers. They include sheathing with copper and other metals, or covering the wood with broad-headed nails, a process known as " scupper-nailing." Such methods are good so long as the protecting surface remains intact and they are generally adequate for dealing with the gribble but even a crack will allow a larval ship-worm to enter and begin boring. Wood can also be impregnated with creosote containing suitable poisons ; certain poison gases have proved excellent, but even this method does not confer indefinite protection. A recent method, which proves unexpectedly efficient for destroying shipworms *after* they have entered the wood, is to detonate charges of explosives close to the infected piles. The animals are killed by the ensuing vibration.

CHAPTER 13

PIER PILES AND FOULING

"In those Regions, the Ships Bottoms are soon covered with the Habitations of Thousands of Animals ; Rocks, Stones, and every Thing lifeless, are covered with them instantly. Even the Branches of living Vegetables that hang into the Water are immediately loaded with the Spawn of different Animals, Shell-fish of various Kinds. And Shell-fish themselves, when they grow impotent and old, become the Basis of new Colonies of Animals, from whose Attacks they can no longer defend themselves."

JOHN ELLIS : *An Essay towards a Natural History of the Corallines* . . . 1755

T H E erection of piers and wharves made of wood, iron or concrete has provided new surfaces for the attachment of the plants and animals of rocky shores. A settlement area has been created which rises vertically from the shallow waters of the sub-littoral zone. The nature and abundance of the attached life varies according to the position of the piles. Those of promenade piers which are built out over exposed sandy shores are usually relatively sparsely covered and so may be those erected in estuaries where there is much fresh water, or in docks where the water is heavily polluted. But where there is good shelter with adequately saline and clean water and especially if there is sufficient tidal movement to bring food to the attached animals and new larvae to replace them as they die, then the piles are often clothed with a dense encrustation of animals and plants which may be easily viewed from a rowing boat at low tide.

A list of the probable inhabitants would be a mere repetition of the names of the commoner animals and plants of rocky shores. Animals tend to predominate because many of the intertidal weeds fail to find adequate surface for attachment. The highest zone is usually occupied by filamentous green seaweeds ; these are then replaced by a dense covering of barnacles which may literally cover the entire surface. These are typically succeeded by a belt of mussels, often so profuse that they grow over one another with the underlying layers smothered by their successors. Nearer low water level come anemones, especially the frilled *Metridium senile,* and these extend into water where they are permanently submerged and in which they grow larger than they do between tide marks. The intertidal zone is occupied also by many other sedentary animals, such as sponges, hydroids, serpulid tube-worms with procumbent limy shells, polyzoans and sea-squirts. Fucoid weeds are usually sparse but there may be a variety of red weeds. Amongst these fixed animals and plants there crawl the crustaceans, worms—especially nereids which abound among the mussels—and snails that scavenge or feed upon them. Wooden piles, unless suitably protected, will almost certainly be at least superficially eroded around low water level by *Limnoria* and *Chelura* (Pl. XXVIIb, p. 174), and old timbers may be riddled with *Teredo*. Below low water level, where the encrusting animal life is not so dense, weeds have a better chance of settling and long tangle-weeds may float upwards and obscure the base of the piles.

At low water of spring tides the white- or orange-coloured masses of dead man's fingers, *Alcyonium digitatum* (Pl. XXVIIIa, p. 175), may be readily seen below the surface and occasionally be exposed. This animal, it may be recalled, is the only example of an alcyonarian coelenterate likely to be encountered on the shore, although it is primarily a sub-littoral species. Out of water the colonies have little to recommend them ; they are but rounded or lobed masses of a tough, jelly-like material strengthened with limy spicules. Through the mass run canals which unite the numerous polyps which expand over the surface when the colony is under water. Each has the eight feathered tentacles typical of this type of coelenterate and the colony has real beauty when the surface is covered with a soft translucent fur of these projecting polyps as shown in Plate XXVIIIa. Animals of this type are among the commonest intertidal inhabitants of Indo-Pacific coral reefs, where wide areas may be covered with dull-green or brown colonies like masses of fleshy seaweed. Another common coelenterate on sheltered pier piles, and one which does not object to some lowering of the salinity, is the hydroid or scyphistoma stage of the common jellyfish, *Aurelia* (Pls. I and II, pp. 16, 17). The jellyfish itself is not infrequently to be seen gently pulsating in the sheltered water around and between the pier piles.

In such still waters the encrusting animals may assume forms unknown on the wave-beaten rocks and may give the initial appearance of different species. Sponges and compound sea-squirts that only extend as films of tissue on the exposed rocks may, where they are not constrained in growth by a rush of waters, extend into long processes or hang down in fleshy lobes from the piles (Pl. XXVIIIb, p. 175). These different growth forms provide excellent examples of the effect of environmental forces. Attached animals cannot avoid the force of the waves by crawling into the protection of crevices or gliding beneath stones. Many survive on the shore only because the body form is plastic and can be moulded by the force of the seas that it must needs resist. All living things represent the result of reaction between inherited capacities on the one hand and environmental factors on the other. The unexpected forms assumed by encrusting animals in still waters reveal the presence of capacities which can never be realised in the turbulent waters on an exposed shore.

Deep shade prevails beneath the planking of a wharf and the population particularly on the inner rows of piles may be enriched by the presence of organisms that normally dwell in the lower light intensity of sub-littoral waters. Exposure in such shelter entails no danger from desiccation or high temperature and red seaweeds in particular tend to extend their vertical range up the surface of these shaded piles. Personal observations on tropical shores revealed similar conditions under a wharf in the Gulf of Mexico where corals were found growing that never live exposed to the full illumination on the surface of reefs.

Consideration of the population of pier piles leads us naturally to the growth of animals and plants on the under-sides of ships, another type of surface presented by man for the attachment of marine organisms. The consequent fouling is a matter of grave economic importance and the desire to overcome it has stimulated much research, especially during the recent war. Some of the results of this work may justifiably be mentioned because they have a wider application to the general problem of the attachment of animals and plants to all manner of hard surfaces, including intertidal rocks.

No sooner does a ship take to the water than plants and animals start to settle upon its submerged surfaces, although this is prevented as far as possible by the use of anti-fouling paints. The speed of fouling depends on latitude and is about twice as rapid in tropical as in temperate seas. Settlement only occurs when the ship is stationary so that ships are most quickly fouled when they lie for long periods in port. There the bottom provides a surface for settlement along with the pier piles and the intertidal rocks. Conditions vary greatly in different ports, fouling being least where there is much freshwater or high pollution.

The effect of such fouling on the speed of a ship is very great. Once the plants and animals have settled they continue to grow while the ship is moving and it has been estimated that in temperate seas the frictional resistance due to the projecting bodies of fouling organisms increases by one-quarter of one per cent per day. Six months after it leaves dock, the maximum speed of a battleship of 35,000 tons is reduced from this cause by $1\frac{1}{2}$ knots while an additional 40 per cent of fuel is needed to maintain a speed of twenty knots. On the average probably 20 per cent of the fuel used by a ship is needed to overcome the added resistance due to the growth of this encrusting life. Frequent lengthy periods in dock for scraping and repainting are an added source of expense.

An enormous increase in the efficiency of naval and mercantile vessels with a great saving of money would be achieved if the growth of this fouling could be prevented or materially reduced. The problem is being tackled in two ways, by the preparation and testing of new types of anti-fouling paints and, the aspect which concerns us here, by studying the nature of the animals and plants involved and of the conditions which control their settlement.

When a clean surface is exposed in the sea a variety of marine organisms quickly accumulate upon it as they do upon any natural surface which has been swept bare. These organisms include bacteria, diatoms, seaweeds and many types of animals. The diatoms are bottom-living representatives of the important group of unicellular plants which form a major part of the plant plankton. Together with the bacteria they form a film of brown slime and sometimes trailing threads over the surface. The seaweeds settle as microscopic spores from which the young plants rapidly grow. They include many forms now familiar to us, including the green *Enteromorpha*, *Cladophora* and, less commonly, *Ulva*. Of brown seaweeds the commonest are probably species of the delicate and much-branched *Ectocarpus*, which form dark-brown masses several inches tall. *Scytosiphon* is the commonest of the larger brown weeds. *Polysiphonia* and *Ceramium* are the most usual red weeds and have the same short tufted growth as the brown *Ectocarpus*.

Animals make contact with the surface in the last stage of planktonic life and almost immediately proceed to change in form and to attach themselves. They include the now familiar sponges, hydroids, of which species of *Tubularia* (Pl. 24a, p. 137) are the commonest in British waters, polyzoans, serpulid tube-worms, barnacles, especially the sub-littoral *Balanus crenatus* (Pl. XIIIa, p. 114), mussels, and compound sea-squirts. With the exception of the hydroids, which are attached by a small area only with the plant-like colony projecting from the surface, all of these animals spread out widely over the surface which in a very short time may become largely covered with the lime or other material they lay down around their bodies for protection and attachment. Even after death in polluted or in fresh waters, the limy skeletons of barnacles and tube-worms persist and are as great a hindrance to movement by the ship as when they were occupied by the animals that formed them.

Plants are naturally most abundant near the water line where they may flourish to the more or less complete exclusion of the animals, which increase in numbers lower down. In the deep shade on the under-side of the projecting bilge keels on ships they form the great bulk of the fouling. Where many plants and animals settle together in numbers the resultant population is very mixed initially but some particular type may settle in overwhelming numbers or find conditions unusually favourable to it so that in subsequent growth it largely swamps its competitors. Similar exclusion of other species has already been noted within the zones occupied by the intertidal fucoid weeds.

A variety of factors influence settlement on the shore and sub-littoral zone as well as on ships. Some have already been noted, namely depth and shade with their effects on the intensity and quality of light, the effect of other organisms competing for space for settlement, and the speed of water movement which was discussed earlier in relation to the settlement of barnacles on the shore. Here it may be noted that there is a difference between the shore-dwelling *Balanus balanoides* and the sub-littoral and common fouling species, *B. crenatus*. Agitation of the water in which the late or cypris larvae are contained causes settlement in the former which normally live in surging seas, but has no such effect on the larvae of the second species which settle in still water offshore.

Seasonal effects and the nature of the surface remain to be considered. Clearly the former will depend on the breeding seasons of the different animals and plants. As shown in Fig. 59 (p. 193), the sea-weeds and diatoms settle throughout the year although most abundantly

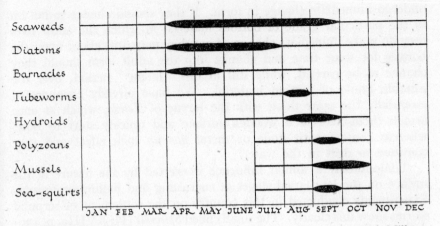

FIG. 59—Seasonal settlement of fouling organisms at Caernarvon and Millport, 1941-42. Periods of heavy settlement shown by thickened, and of light settlement by thinner lines. (Adapted from *Fouling of Ships' Bottoms: Identification of Marine Growths*, The Iron and Steel Institute, Marine Corrosion Sub-Committee, 1944.)

in the spring and summer. The animals only settle within restricted periods, namely at the end of the freely swimming larval period of the life-history when they are members of the temporary plankton. There is thus a gap, dependent on the length of this planktonic period, between the time of spawning by the parent and of settlement by the young.

The barnacle *B. balanoides* is the first species of animal to settle in the year, settlement extending from the beginning of March to the end of August with a peak period in April and May. *B. crenatus* settles later, reaching its maximum in June. Serpulid tube-worms start and finish about a month later but settle most thickly during August. Hydroids attach themselves from June to October and especially in August and September, while polyzoans, mussels and sea-squirts do not begin to settle until July and then continue through to October with the peak period largely in September. These periods of settlement are those that prevail off our western coasts ; similar animals tend to settle somewhat earlier in the south and later in the east and north.

The nature of the surface has an important effect which would appear to be exerted upon the later larvae when due for settlement. It is known that when larvae of a variety of bottom-living animals sink to the bottom at the end of planktonic life they do not necessarily undergo immediate change in form. If they should chance to fall on to the particular grade of bottom material in which the adult lives they do make immediate change, if not they may drift about upon the bottom for some time and change into the adult form should they chance to be carried, while still in this " latent " period, over the suitable grade of bottom material. We have already encountered essentially the same thing with the larvae of *Teredo*, which at once attach themselves to a wooden surface and quickly start to bore, whereas contact with stone or metal has no such effect and they continue to drift in the water.

Undoubtedly a similar influence is exerted by the nature of the surface on the late larval stages of encrusting and fouling organisms. Experiments conducted in this country on the settlement of serpulid worms show this clearly. The small coiled *Spirorbis* (Pl. XVIIIa, p. 139) and the larger *Pomatoceros* (Pl. XIXa, p. 142) with keeled tube both settle almost exclusively on smooth surfaces while the intermediate-sized *Hydroides*, which often grows in large masses, accumulates with equal speed on smooth and on rough and granular surfaces. In the former cases apparently only contact with a smooth surface causes the larvae to change form and become attached but those of *Hydroides* are accommodated more easily. Experiments in America in which a variety of materials were exposed in sea water showed that, in general, organisms accumulated more rapidly on porous or fibrous surfaces. A wide field for observation and experimentation is here presented and the results may well contribute towards the solution of the problem of fouling. It is certainly only in this way that we can hope to solve the purely scientific problem of the nature of the factors which control the distribution of seaweeds and animals which attach themselves not only to pier piles and to ships but also to the rocks of the sea shore.

CHAPTER 14

ZONATION ON ROCKY SHORES

"Involved in sea-wrack, here you find a race,
Which Science, doubting, knows not where to place ;
On shell or stone is dropp'd the embryo seed,
And quickly vegetates a vital breed."

GEORGE CRABBE

THE zoning of both plants and animals within the narrow vertical limits of a rocky shore is one of the most striking features about shore life. Long and arduous ascent from sea level to mountain top is necessary to observe its counterpart on land. Many instances of zoning have been mentioned in preceding chapters but the subject is far from exhausted and will repay further discussion before we pass on to the description of other types of shore. There, it is true, we shall find zonation of life in sand and, for somewhat different reasons, up estuaries, but not to the same extent or so obviously as on rocky shores. Only on these grow the successive bands of fucoid seaweeds while the zonation of barnacles, periwinkles and top-shells has no obvious counterpart in the concealed life of sandy and muddy shores.

The cause of zonation resides in the varying extent to which different plants and animals are adapted for life on the shore. The more highly adapted the higher do they live upon it. There is, however, more to the matter than this because as they reach upward towards high water level they cease first to inhabit the sub-littoral zone and then the lower regions of the shore itself. This is but one out of innumerable instances of the general truth that adaptation to a new mode of life involves loss as well as gain. The plant or animal gradually becomes confined to the new environment which its powers of adaptation have enabled it to conquer. The necessary changes in form and in function may lead the organism to a dead end ; such is the fate of parasites, supremely successful although they frequently are within their particular sphere of life. The same may be said of the acorn-barnacles that fringe the shore ; fixed to the rock and yet dependent upon the sea for food and as a medium for early freely-swimming life, they cannot retreat further from it. But to mobile animals the upward ascent of the shore may lead, as it has undoubtedly led in the past, to eventual conquest of the vast new environment of the dry land. Snails, isopod crustaceans and crabs, all contain terrestrial representatives that must, by progressive stages of adaptation for life on higher and yet higher zones on the shore, have in past time at last acquired the capacity to extend beyond its limits on to the surface of the land. Some existing members of our shore fauna show clear evidence of such tendencies.

The competition for space on the shore is so intense that only the particular species of animals and plants most perfectly adapted for life in any zone, or in any special habitat within a zone, are able to maintain themselves. This becomes apparent when an area of rock is artificially bared ; the initial population which recolonises it is usually mixed and more variable than the final population. Beyond possibly providing a substratum for them, as an initial covering of barnacles may do for subsequent growths of weed, those that come first do not assist the settlement of their successors whose final success represents the survival of the most fitted.

There is displayed diagrammatically in Fig. 60 (p. 197), the vertical zonation of the commoner brown and red weeds and animals on the south coast of Devonshire. All of these species have already been described at some length. We can best begin with the plants, which range so beautifully from the maximum exposure of *Pelvetia* to the rare uncovering of *Laminaria* at the lowest ebbs. Motility in these is limited to their brief period as free spores which quickly settle to attach themselves as sporelings on the surface of rocks. But however quickly they settle they are, for an all-important period, at the mercy of complicated tidal movements and only a very small proportion can ever settle within the limited zone where adult life is possible. The spores are scattered haphazard and only if their inherent properties fit them for adult life in the area where they chance to settle can the sporelings grow to maturity.

The zonation of the intertidal fucoid weeds appears to be initially controlled by two factors, their capacity to withstand desiccation and the speed with which they grow. On the upper shore it is the former which matters but on the less exposed middle and lower shores success goes to species that can outstrip their competitors in speed of growth. The capacity to withstand water movements and, especially low on the shore and in the sub-littoral zone, to live in a region of low light

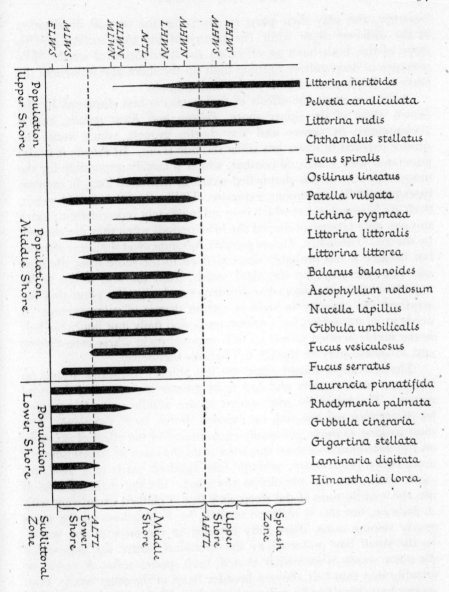

FIG. 60—Zonation of intertidal animals and plants on the landward portion of Church Reef, Wembury, near Plymouth. (Adapted with modifications from Evans, *Journal Marine Biological Association*, Vol. XXVII, p. 209.)

intensity, also play their parts in governing the vertical distribution of the different algae while the nature of the rock surface and its slope of this both have an effect on the distribution of weeds. The presence of deep gulleys running high up the shore also influences the vertical distribution of both plants and animals.

Ability to resist the effects of desiccation is best displayed by the brown algae of the upper shore, *Pelvetia* and *Fucus spiralis*, and is accompanied by slower and also shorter growth, which limits the surface exposed to water loss when the tide is out. In *Pelvetia* it is also associated with a high oil content, which is largely responsible for the unique capacity of this channelled wrack to withstand long, in extreme cases practically continuous, exposure. Unlike the other fucoid weeds, both are hermaphrodite which may aid in efficient reproduction ; since this can only take place during the brief periods when they are covered by the sea. The lichen, *Lichina pygmaea*, has somewhat similar properties but because it has probably descended on to the shore from the land and not emerged, like the algal weeds, from the sea, it has to be considered from the somewhat different standpoint of a plant that has acquired the capacity to resist a certain degree of submergence. It needs to be uncovered for a minimum period daily and flourishes best in the highly aerated waters on fully exposed rocks where even *Pelvetia* and *F. spiralis* may be unable to establish themselves.

The remaining fucoid algae are less able to withstand drying up but grow more rapidly and can resist greater movement of the tidal waters. The relatively wide extent of the middle shore is occupied by *Ascophyllum nodosum* and, on exposed shores, by *F. vesiculosus*. The most striking example personally encountered of the effect of exposure on the distribution of these species was on the coast of Argyll where a short rocky promontory, perhaps two hundred yards long and fifty yards wide, runs out parallel to the coast. On the sheltered side of this the middle zone of the steep rock face is clothed exclusively with *A. nodosum*, but this is replaced entirely by *F. vesiculosus* on the more gently sloping outer shore fully exposed to the force of the waves. In the small bay protected by the headland, where the violence of the outer waters is somewhat abated, both species occur, *F. vesiculosus* growing in a thin belt above a broader band of the other weed. Both species have long fronds and are buoyed by bladders but *F. vesiculosus* has greater tensile strength and is thus able to resist more powerful seas, while the even greater length and the elasticity of the fronds of

Ascophyllum render it admirably fitted for life in the up-and-down swell and surge of more sheltered waters.

The serrated wrack, *F. serratus*, of the lower shore has somewhat shorter fronds and even if buoyed they could not reach the surface during high tide as do those of the bladdered wracks. It has very restricted powers of withstanding desiccation but can flourish in a somewhat lower light intensity than the species found higher on the shore. In texture it is not so elastic as *Ascophyllum* but possesses greater tensile strength. There is less to say about the two red algae, *Gigartina* and *Chondrus*, which inhabit much the same area. They are too short to be significantly affected by water movements and it is their limited powers of withstanding desiccation together with their preference for a particular intensity of illumination which probably control their vertical distribution, although their abundance on any stretch of shore appears to depend upon the slope and nature of the rock surface. The large brown weeds which fringe the upper margin of the sub-littoral zone, *L. digitata*, *Himanthalia*, and, in certain areas, *Alaria*, all possess pliable stems and so are the better able to withstand both violent wave action and also, by lying prostrate, exposure to desiccation when the tide is out.

The important fact has been pointed out by Mr. R. Elmhirst that the six-hour period when the tide is rising or falling may be roughly divided into three. During the first and third of these the tide covers or uncovers about one-quarter of the shore (the actual distance varies according to the state of the tides) and during the second period rises or falls approximately twice as quickly. This is indicated in the curves of tidal movement in Fig. 23 (p. 60). At spring tides when the full extent of the shore is covered and uncovered, the water will thus advance or retreat slowly over the lower and upper shores and quickly over the middle shore. During the lowest neaps only the middle shore will be affected ; but always the fucoids of this region will experience greater tidal movements and also the more frequent shock of breaking surf than those of the upper shore. The serrated wrack, although submerged on the lower shore during the periods of greatest tidal flow, has to counter the drag of the water at these times and for this it is well adapted owing to its great tensile strength. Where wave action is reduced it extends significantly higher up the shore than on exposed coasts.

The algal inhabitants of the three major regions of the shore are thus beautifully adapted for life within the limited extent of each zone, which may be relatively broad on a gently sloping shore or shrink to a few feet on a steep rock face. On overhanging surfaces, it may be noted, they fail to establish themselves, and there barnacles may exist alone. Similar powers of adaptation explain the similar restricted distribution of other species of brown, as well as of green and red, weeds. One cannot too often emphasise the widely different conditions for life that exist within the narrow confines of the shore. Immobile organisms in particular, and they include all the seaweeds, can only exist within a certain range of these conditions and this is reflected in the restricted and sharply defined vertical zones to which each species is confined. Other factors may influence the horizontal distribution but as these also affect the animals they will be discussed a little later.

On typical rocky shores in this country the dense growth of successive bands of seaweeds obscures the zoning of the animals. We habitually refer to the zones in terms of the algal population, *Pelvetia* zone, *Ascophyllum* zone and so forth. But these intertidal fucoids, together with the sub-littoral tangles, are confined to temperate and cold seas and are absent in the tropics and on the shores of warm temperate lands. There the dominant shore inhabitants are animals and it is their zonation which is immediately obvious. Such conditions prevail, for instance, along the shores of South Africa, the fauna and flora of which have probably been more fully described than those of any other region of comparable size during the course of a ten-year survey carried out by Professor T. A. Stephenson and colleagues from the University of Cape Town. This revealed the presence of three major zones from above downwards, a *Littorina* zone extending from above the limits of the true shore, in which species of this snail were dominant, a *Balanoid* zone characterised by the presence of acorn-barnacles and the limpet, *Patella granularia*, which sometimes occurs alone, and below this a *sub-littoral fringe* including the lower shore and the beginning of the sub-littoral zone where there is no one dominant animal. In some areas an additional zone, characterised by another species of limpet, was identified above the sub-littoral fringe.

There is evidence that the zoning of animal life follows essentially the same pattern on our own shores. In regions of great exposure or where the rock face is steep the seaweeds may be reduced to isolated patches or even be absent. Under such conditions what may be regarded as the basal zonation of animal life is laid bare. There is a high zone occupied by the small and rough periwinkles, *Littorina*

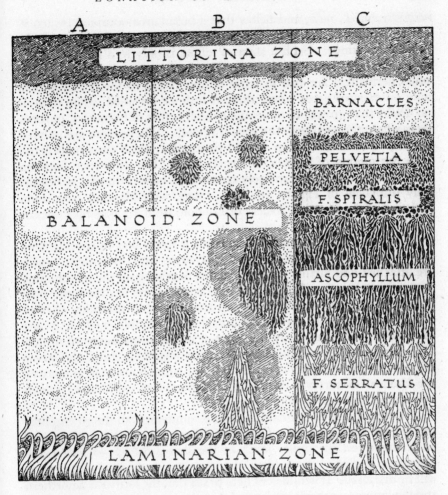

FIG. 61—Diagram illustrating three of the possible types of zonation found on British shores (e.g. on steep rock-faces in Argyllshire). Only part of the width of the *Littorina* and *Laminaria* zones is included. The relative widths of the other zones are accurate for the places chosen. In column B, areas of bare rock (more or less free from barnacles) surround the larger fucoid plants. These areas are due to the movements of the algae when submerged. For column C a locality without *Fucus vesiculosus* or *Porphyra* was chosen. For further explanation see text. (Reproduced from Stephenson, *Journal Linnean Society, Zoology*, Vol. XL, p. 508, by kind permission of the author and of the Linnean Society of London.)

neritoides and *L. rudis*, and below that a broad area occupied by acorn-barnacles and bounded on the seaward side by the *Laminaria* of the sub-littoral fringe. Such a state of affairs is only to a limited degree possible where seaweeds abound. The freely swimming larvae of the small periwinkle are said to settle in abundance in the barnacle zone only on exposed areas where there is little or no weed, while there is no doubt that growth of these algae inhibits that of barnacles, particularly where the surface of the rock is smooth. Where there are isolated patches of weed the region around them, over which the fronds brush with the motion of the sea, is largely bare. The difference between these shores, densely or sparsely covered or bare of weeds, is beautifully displayed in Fig. 61 (p. 201), taken, with his kind permission and that of the Linnean Society of London, from a paper by Professor Stephenson.

In discussing the zonation of animals it seems best to consider first those, such as barnacles, sponges, polyzoans and sea-squirts, which are attached and further resemble plants in relying on the surrounding water for food, though they also live freely in the sea for some time before attachment. Chance plays a major part in deciding whether a particular larva is carried to a region of possible settlement and inevitably by far and away the greatest numbers are borne to unsuitable areas. But once safely deposited upon a suitable shore, the larvae of barnacles and possibly of other attached animals are able to increase the chances of survival during a brief period of exploratory search before final settlement (Fig. 33, p. 113).

Of these attached animals much the commonest and most carefully studied are the barnacles. Two species have already been mentioned as occurring on the upper levels of the shore, *Balanus balanoides* and *Chthamalus stellatus*. In a later chapter their horizontal distribution will be discussed. It is sufficient here to say that the former is a northern and the latter a southern species and that they overlap along our western shores. Because they are members of different faunas, the late larvae or " spat " of *B. balanoides* settle in the early summer but those of *Chthamalus* not until the autumn and winter. This gives the former the better chance to become well established before the winter storms begin. The upper limit of *Balanus* is almost along the level of high water of neap tides, while *Chthamalus* goes higher (see Fig. 60, p. 197), up to and even beyond extreme high water level of spring tides ; the two may overlap although it is uncertain whether there is any real competition for space or food. There are regions, however, where their zones are separated and then the lower level of *Chthamalus* cannot be the result of successful competition by the other species. It has been ascribed to the effects of excessive immersion—as though this typically shore-living animal may be drowned if kept too long under water.

A very interesting instance of the effect of the nature and texture of the substratum on the settlement of barnacles has been described by Dr. H. B. Moore and Dr. J. A. Kitching, to whom much of our knowledge about the distribution of *Chthamalus* is due. On the cliffs below Tilly Whim caves near Swanage the rock strata, of limestone and chert, are horizontal, each with a characteristic texture at the vertical weathered face. Full details are given in Fig. 62, from which it will be seen that *Chthamalus*, as always, occurs above *Balanus*,

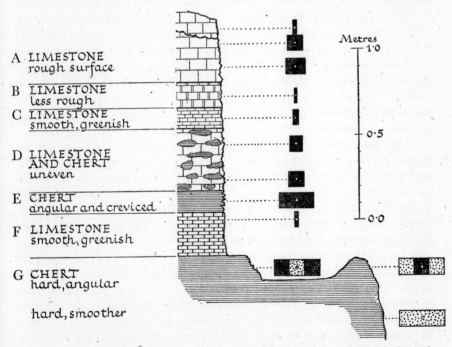

A LIMESTONE
rough surface

B LIMESTONE
less rough

C LIMESTONE
smooth, greenish

D LIMESTONE
AND CHERT
uneven

E CHERT
angular and creviced

F LIMESTONE
smooth, greenish

G CHERT
hard, angular

hard, smoother

Metres
1·0

0·5

0·0

FIG. 62—Vertical distribution of the acorn-barnacles, *Chthamalus stellatus* (black) and *Balanus balanoides* (stippled) in relation to nature of exposed strata of cliff-face at Tilly Whim, near Swanage, Dorset. Brick pattern indicates limestone and horizontal shading indicates chert. (After Moore and Kitching, *Journal Marine Biological Association*, Vol. XXIII, p. 533.)

the two overlapping to some extent, but that the population varies on the different bands of exposed rock. On shores where the rock surface is uniform, the numbers of barnacles increase downward, reach a maximum and then decline in lower levels. But here, within the limits of settlement, abundance is clearly related to the nature of the surface. The heaviest populations are carried on bands A, E and G which, especially the two last, are hard with a coarse surface. Band D, which carries a good population, has hard nodules of chert among the limestone. In the remaining strata where the surface is smooth the barnacles are found in much smaller numbers, probably because in the absence of protecting crannies the larvae are carried away by water movements before they are able to settle.

The zonation of barnacles, and here we may also bring into consideration *B. perforatus*, which extends upward from the sub-littoral zone to about the middle of the shore, would therefore appear to be controlled at its upper limits by the capacity to withstand prolonged exposure, best developed in *Chthamalus* and least in *B. perforatus*, and at its lower limits by the effects of immersion, without influence on *B. perforatus* but certainly harmful to both of the other two species though most to *Chthamalus*. Competition between the species seems to have little influence on zonation, but an unsuitable substratum may reduce settlement.

The same rare chance of successful settlement is the fate of the planktonic larvae of motile shore animals such as limpets, top-shells and periwinkles, with which we may also include the mussels, motile in early life and always capable of movement. Once larvae settle upon a suitable shore, however, the activity of these young molluscs gives them an added opportunity of moving to the appropriate zone. The small periwinkle after settlement in the barnacle zone must always seek its adult home in this way. Where the planktonic stage is omitted from the life-history, as in the rough and flat periwinkles, in the dog-whelk and in the little bivalve, *Lasaea*, the young emerge on the appropriate zone. It is true that the young dog-whelks are carried initially further down shore but they return from thence after a period of growth. The latter method of development is safer but the retention of the planktonic stage gives greater opportunity for wide dispersal of the species.

Attached animals and plants have the advantage of not being displaced except during the heaviest seas. Motile animals, as we have seen, display a variety of devices for resisting the force of the sea, like limpets, or avoiding it like periwinkles and dog-whelks by sheltering in crevices or under stones. But inevitably many must be rolled about the shore during storms and be carried above or below their normal zone. Here behaviour comes into the picture again. With the return of calm weather, the animals proceed to move back, up or down the shore as the need may be, to the area from whence they were carried. Scatter a mixed collection of different species of periwinkles up and down the shore and they will slowly but surely sort themselves out.

Analysis of the behaviour of the three intertidal periwinkles reveals that their position on the shore is controlled by the external factors of temperature, of gravity, influenced in its turn by desiccation and immersion, and of light. Given the choice, the rough and common periwinkles, which live either higher on the shore or more exposed than the flat periwinkle, " prefer " somewhat higher temperatures than that species. Periwinkles higher on the shore show a greater tendency to move upward against the force of gravity than do those lower on the shore, the former displaying, to express the matter technically, greater negative geotactic response. The effect of this reaction will be to cause the animals to move up the shore and as the average speed of a common periwinkle is some four feet per hour they can readily move from one level to another. But this upward movement is checked by the effect of desiccation, which if prolonged beyond a certain point causes the negative geotaxis to be reduced or lost. Prolonged immersion has the opposite effect, causing the animal to move upshore. Hence above a certain degree of exposure the tendency to move higher is checked or even reversed, above a certain degree of immersion it is increased. The combined effect of these external factors is to drive the members of a particular species within certain zonal limits characteristic of that species.

The effect of light is somewhat similar although it varies more between the three species. The rough periwinkle, which lives exposed on rock faces, tends to move towards the light ; it shows, as we say, positive phototaxis. Hence it keeps in the open. This reaction to light is less marked in the common periwinkle, which lives sometimes exposed and sometimes in shelter under stones or in pools, while the flat periwinkle, which always lives in shelter among weeds or under stones, moves away from light, or in technical terms is negatively phototactic. These reactions to light condition the habits of the different species within the zones to which they are restricted by their reactions to the force of gravity.

The behaviour of other intertidal molluscs can be similarly explained. One may assume negative geotaxis in the young of the small periwinkle when they move up to the splash zone after settlement on the upper shore ; indeed when placed in a bowl of water these animals always crawl to the top. This reaction has the important consequence of taking them beyond the action of the heavy surf on the exposed coasts that they frequent. Only extreme exposure checks the upward movement. After initial growth at low levels negative geotaxis would account for the upward movement of young dog-whelks to the zone where they were hatched. The sometimes very restricted vertical range of the top-shell, *Osilinus lineatus* (Fig. 60, p. 197), may be attributed to the delicacy with which the reactions to gravity are influenced by increased exposure on the upper and increased immersion on the lower limit of its zone.

Survival in the particular zone occupied by any of these molluscs depends on the efficiency with which the animal can resist desiccation. The various means employed to this end have been described, but it is worth while referring for a moment to work on the large whelk, *Buccinum undatum*. This is a sub-littoral species only occasionally, and only when young, found on our shores and then usually in well-sheltered areas. But in the Bay of Fundy, on the Atlantic coast of Canada, where the tidal range is the greatest in the world, numbers of these whelks tend to move up to low water level during the period of neap tides so that they are left well exposed on the lower shore during the great ebb of the succeeding spring tides. They thus become, for the time being, shore-living animals but without the appropriate reactions and protective devices. Instead of withdrawing within the shelter of the shell and closing the opening with the operculum, like a periwinkle or a top-shell, they continue to crawl about and quickly lose water from the gill cavity and the tissues. Movements are purely haphazard and by chance may carry a few whelks to shelter or back into the sea. But inevitably the majority of these exposed animals die from the effects of desiccation. This is of no serious consequence to the species as a whole because the great bulk of the population remains in the sub-littoral zone. From the failure of these whelks to survive even brief exposure on the shore we can appreciate both the rigours this represents and the true significance of the behaviour and protective devices which enable the true shore-living animals to maintain themselves, each within its particular zone, between tide marks.

With the exception of those few animals that hatch out on the shore, zonation is thus the result of the settlement and survival to adult life of some minute proportion of the countless numbers of spores and larvae thrown by the sea upon the rocks. The force of the seas on the most exposed shores permits the settlement of little more than barnacles and the small periwinkle. With some increase in shelter come the dog-whelks which feed on the barnacles. Where water movement is not too violent, algal spores are able to attach themselves and, as Mr. R. G. Evans has recently pointed out, there is an algal invasion from both the upper and the lower shores. *Pelvetia* and *Fucus spiralis* descend from above, while there ascend, in order from below upward, *Laminaria digitata*, *Corallina*, *Gigartina* and *Laurencia*, followed by *F. serratus* and then *F. vesiculosus*, which is replaced by *A. nodosum* in more sheltered waters. In consequence the continuous intertidal zone of balanoids, shown in Fig. 61A (p. 201), is obliterated and replaced by the successive bands of seaweeds shown in Fig. 61C. With these weeds come the animals to which they afford protection and food, namely the common and flat periwinkles, limpets, top-shells, worms and crustaceans, some confined to restricted zones, others, such as the common shore-crab, ranging at will over the shore.

Where there are pools, zonation, as we saw in an earlier chapter, is controlled by factors other than desiccation, namely by the varying capacity of plants and animals to withstand wide and often sudden variations in temperature, salinity and acidity. Conditions vary widely according to the nature and situation of the pools. The population of shallow pools on the exposed surface of rocks is affected by water movements while that of pools around the outflow of streams is limited by the abnormal amounts of fresh water. Seasonal effects are well marked ; crustaceans and shore-fishes migrate upshore in summer and may retreat into sub-littoral waters in the winter, the algal population ebbs and flows, some weeds moving higher in the summer and others in the winter.

If rocky shores had everywhere the same slope and were exposed to the same forces from the sea their population would not vary from place to place ; an uninteresting uniformity would prevail. In nature this is far from being the case. The population of a precipitous rock face consists largely of algae and attached animals but even the former

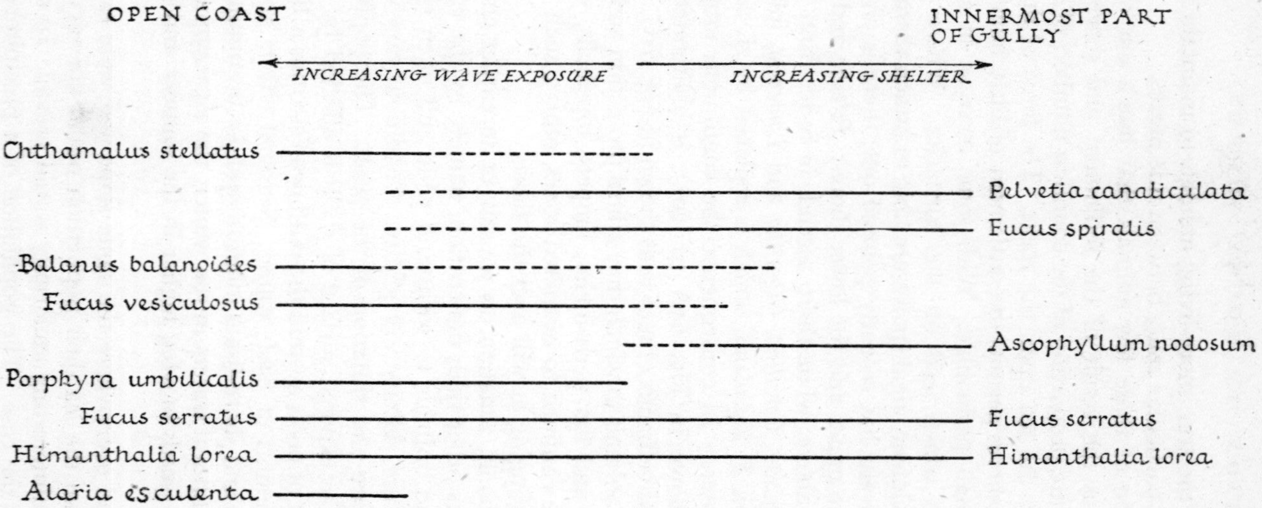

OPEN COAST   INNERMOST PART OF GULLY

← INCREASING WAVE EXPOSURE   INCREASING SHELTER →

Chthamalus stellatus

Pelvetia canaliculata

Fucus spiralis

Balanus balanoides

Fucus vesiculosus

Ascophyllum nodosum

Porphyra umbilicalis

Fucus serratus

Fucus serratus

Himanthalia lorea

Himanthalia lorea

Alaria esculenta

FIG. 63.—Diagram showing distribution of various species of animals and plants with relation to varying degrees of wave-exposure, in a gully at the south-west end of Goat Island, the Small Isles, Jura. (After Kitching, *Transactions Royal Society of Edinburgh*, Vol. LVIII, p. 368.)

cannot exist on overhanging surfaces, probably because the spores do not settle under such conditions. The effect of surface texture on attachment was well shown in the case of the barnacle, *Chthamalus* (p. 203). Extreme exposure prevents the settlement of algae and of all but a few animals. But wherever the shore is sufficiently exposed to steady pounding by the waves there will be a splash zone, in some cases as wide or wider than the true shore, and the effective vertical extent of the shore is correspondingly increased. The population of such shores extends higher than that of protected shores. The effect of shelter on the population of encrusting barnacles and algae is best shown by a comparison between their distribution on the exposed and sheltered ends of a rocky gully, such as one studied by Dr. J. A. Kitching on a small island in the Sound of Jura off the west of Scotland, the results of which are summarised in Fig. 63. The barnacles, *B. balanoides* and *Chthamalus*, are essentially creatures of exposed waters with the latter, here and typically, extending a little further into shelter. Of the seaweeds, the sub-littoral tangle, *Alaria*, is revealed as the most dependent on wave action, but *Porphyra* and, of course, *Fucus vesiculosus* disappear as water movement diminishes within the shelter of the gully. The algae of the lower shore, the serrated wrack and *Himanthalia*, appear to be unaffected by the degree of exposure but the typical algae of the upper shore, *Pelvetia* and *F. spiralis*, are absent on the most exposed shores. *Ascophyllum* we already know to be a plant of sheltered waters and is here completely replaced by *F. vesiculosus* when the degree of exposure rises above a certain point. Observations in other regions, as far apart as the Faeroes and Brittany, do not always indicate the same order of susceptibility of these algae to exposure. Possibly the texture of the rock surface influences settlement of algae as it does of barnacles : some weeds may be better able to maintain themselves in regions of greater water movement if the rock is rough than if it is smooth.

The effect of exposure on the vertical distribution of the intertidal flora and fauna may now suitably be summarised. Extreme exposure inhibits the growth of plants, but first the lichen, *Lichina pygmaea* and then the algal weeds appear as the force of the surf abates. The effect of splash enables *Pelvetia* and *F. spiralis* to extend higher than on sheltered areas, whereas the reverse is true of *F. serratus*. There are also indications that the level of *Laminaria* and of *F. serratus* and *F. vesiculosus* is lower in the south than in the north of Britain. Among

T.S.S.   P

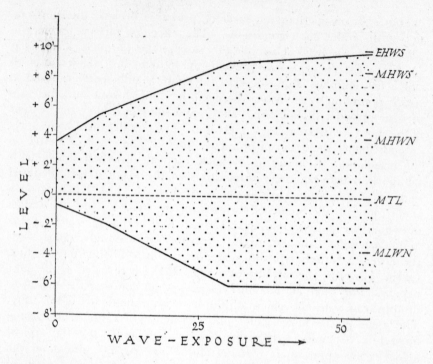

FIG. 64—Variation in the upper and lower limits of distribution of the acorn-barnacle, *Balanus balanoides*, with varying degree of wave-exposure. (From Moore, *Journal Marine Biological Association*, Vol. XX, p. 282.)

animals, increased exposure causes both upward and downward extension on the shore of *B. balanoides* (Fig. 64). The former effect is probably due to greater transport of food and oxygen and to the fact that the settling larvae may be washed higher ; the latter may be the result of greater aeration in more turbulent waters. The dog-whelk accompanies its prey in both directions and the small periwinkle similarly widens its range, extending far into the splash zone above. Upward movement alone occurs in *Chthamalus* and *Balanus perforatus* and in the rough and common periwinkles, in some top-shells and in the species of *Patella*. It has been found that dense settlement of the larvae of *P. vulgata* occurs in heavy surf. As we have seen, disappearance of the weeds is accompanied by that of the various animals that live upon or in them.

Various attempts have been made to establish the presence of a series of critical levels which determine the vertical distribution of intertidal plants and animals. At best such levels cannot be absolute : they must be influenced by the degree of exposure and of slope and by other factors, possibly unsuspected, because we have still much to learn about the interplay of forces on the shore. All I feel that we can be certain about at present is the significance of the division, suggested in Chapter 5, into upper, middle and lower shores. The suggested boundaries were the upper and lower average tidal levels which, as seen in Fig. 23 (p. 60), lie respectively midway between MHWS and MHWN and between MLWN and MLWS. Now, as Mr. R. G. Evans, in particular, has pointed out, the latter level does correspond very closely to the seaward extension of the majority of intertidal organisms (see Fig. 60, p. 197) while MLWN represents the upper level to which various essentially sub-littoral species—from Professor Stephenson's sub-littoral fringe (Fig. 61, p. 201)—can penetrate. The upper level of significance, which Mr. Evans places between LHWN and MHWS, i.e. very little lower than what is suggested here, again corresponds to that at which the true intertidal animals cease, although extending higher where there is much exposure and the effective region of the middle shore is thereby increased. Thus the middle shore, within the average tidal range, is the least critical area because conditions there prevailing vary little ; there is neither the prolonged exposure of the upper shore nor the equally prolonged submergence of the lower shore. In short, the true intertidal, or shore-dwelling, population consists of species adapted for regularly alternating periods of exposure and submergence with the ebbing and flowing of the tides.

Some of the inhabitants of the middle shore extend down on to the lower shore to mingle with the advancing fringe of the sub-littoral population. Those of the upper shore consist in part of species, such as *Chthamalus stellatus* and *Littorina rudis*, which extend far up from the middle shore, and in part of a distinct population of highly specialised species essentially confined to this area. These include the seaweed, *Pelvetia*, the small periwinkle, *L. neritoides*, and also crustaceans such as the sea-slater, *Ligia oceanica*, and the amphipod hoppers. With increased powers of withstanding desiccation, these have largely or entirely abandoned the middle shore.

In Fig. 60, the vertical distribution of the commoner shore-living species is shown together with the critical levels represented by the boundaries between the three regions of the shore. Vertical lines separate the populations of the three regions. The six species on the right are members of the sub-littoral population which extend over the lower shore and, in the case of the red algae, *Laurencia* and *Rhodymenia*, well on to the middle shore. The central block consists of the inhabitants of the middle shore—the characteristic intertidal species—some of which extend down on to the lower shore but essentially none on to the upper shore. Finally on the left are inhabitants of the upper shore, consisting of the two groups already mentioned.

The shore represents one of the avenues, estuaries being the other, whereby marine animals have passed to eventual conquest of the land. This may be the final outcome of progressive specialisation for life successively on the lower, middle and upper shores, and a brief discussion of this theme forms a natural conclusion to this chapter. Two quite distinct matters are involved, first the capacity of the adult animal to live out of water and second its ability to reproduce itself under purely terrestrial conditions. It by no means follows that both requisites for land life are acquired together.

An admirable instance of this is provided by the two periwinkles that live highest on the shore. *L. neritoides* is almost a terrestrial animal, often living where it may seldom be wetted even by spray. It has survived out of water for up to five months. The gill chamber is modified to act as a lung although not to the same degree as in the true land-snails. Nevertheless it remains dependent on the sea for reproduction, having planktonic larvae like the common periwinkle of the middle shore. *L. rudis*, although not extending quite so high, is in a better position for eventual colonisation of the land. The animal is not so highly adapted for breathing air or for resisting desiccation but it *is* viviparous and so potentially independent of the sea for reproduction. With relatively minor added powers it could live and reproduce itself on dry land.

Other inhabitants of the upper shore which have practically emancipated themselves from dependence on the sea are various crustacea, including the sea-slater, *Ligia oceanica*, and various amphipods such as the sand-hopper, *Talitrus saltator*, which is described in the next chapter. The sea-slater, as we have seen, lives at or above high water level and comes out at night to scavenge the shore at low tide just as its close relatives, the woodlice, emerge from shelter to feed on decaying vegetation and other organic debris. Like all isopod, and amphipod, crustaceans, it is independent of the sea for reproduction, carrying the developing eggs in a brood pouch from which the young emerge in the adult form. The terrestrial woodlice presumably passed through a period of similar life around high tide levels before finally breaking association with the shore and exploiting, with a success revealed by their present ubiquity and numbers, the resources of the land. Such passage involves no modification of the pre-existing means of locomotion, feeding or reproduction ; only a limited added capacity for resisting desiccation and for breathing air. To this day woodlice remain incapable of surviving in a dry atmosphere. The culmination of adaptation for life on progressively higher zones on the shore may thus be the final severing of contact with the sea and the transformation of a marine into a terrestrial animal.

CHAPTER 15

LIFE ON SANDY SHORES

"Above all there was Whitsand Bay. . . . There was a rough path leading to an exquisite beach of white sand, over which curled and dashed waves from the Atlantic, bringing in razor shells, tellinas of delicate pink, cockles, and mactras. It was the most delicious place that I ever knew, and to this hour a windy night will make me dream of the roll and dash of its waves and the delight of those sands."

CHARLOTTE MARY YONGE : *Fragments of autobiography published in 1892 and quoted by Georgina Battiscombe.* 1943

THE appearance of a sandy shore as it is uncovered by the falling tide is in startling contrast to that of an exposed rocky shore. In place of the rich covering of seaweeds with the numerous animals exposed on the rocks or sheltered in the pools, the sand appears barren and without trace of life. After it has been bared for a while the smooth surface may, it is true, become broken with worm casts and depressions of various kinds, indicative of the presence of animals below. But never more than a small proportion of this buried life gives even such indirect evidence of its presence.

Existence upon and within sand makes very different demands on adaptation from life on rocks and in consequence the population is totally distinct. It is also much more restricted in wealth of species, although, under favourable conditions, some of the animals may occur in immense numbers. A sandy shore provides no surface for attachment so that seaweeds with their rich associated fauna, and sessile animals such as barnacles, hydroids, sponges, polyzoans and sea-squirts, are absent, together with many other animals which can only move on a hard surface, like many snails. Those types of crabs, worms and other creatures which lurk in the shelter of rocks when the tide is out or the waves are beating on the shore find no such protection on the flat expanse of an exposed sandy shore. Protection from the force of pounding breakers can there only be attained by burrowing and such powers are essential for life in sand. Once acquired, however, this burrowing habit has the added advantage of enabling its possessors to escape the effects of desiccation so that this factor, of such supreme importance to the dwellers on rocky shores, plays a very minor part in the control of life on, and in, sand.

Before proceeding further we should first inquire a little more fully into the nature of this sandy medium. In an earlier chapter reference was made to the durability of sand, a result of survival in transport of only the harder rock fragments and of their small size which enables each grain to be separated by a film of water from its fellows. This property of capillarity enables sand to retain water long after the tide has fallen to lower levels. Thus it is that, except high on the shore where the sea extends only at the highest spring tides, the sand remains damp and the danger of desiccation is removed. Even when strong winds or a powerful sun are drying the surface this has only a superficial effect which the animals counter by retreating a little deeper.

The effects of both fresh water and exposure to air temperatures are much less than on rocky shores. Streams of fresh water often run over sandy shores, on to which rain may also pour when the tide is out. But such fresh water has surprisingly little influence on the salinity within the sand ; even when it is running over the surface the salinity of the water in the sand is unaffected below a depth of at most ten inches. The animals have merely to burrow below this level to escape the effects of lowered salinity. The midday heat of a summer sun may raise the surface temperature of the sand until it is as much as 10° C. higher than that of the returning sea. But again this rise in temperature, or corresponding fall in winter, affects only the uppermost levels of the sand and below a depth of some six inches the temperature remains almost constant. A difference of 7° C. has been measured between the temperature at the surface and at this depth. Sand-living animals are thus largely immune from the effects of even extensive surface changes in salinity and temperature, because burrowing leads them into depths where conditions remain almost constant. Where the animals of a rocky shore confront or circumvent danger, those of a sandy shore burrow to avoid it ; they truly stoop to conquer. Only on the most exposed Atlantic beaches, where the sand is churned by the surf, may they fail to establish themselves.

Sands vary in the nature of their constituent particles but this is largely immaterial ; of more significance to the contained animals are the differences in size or grade of these particles. Within usual limits this has little effect upon the water content of the sand. This ranges around 44 volumes in 100 volumes of wet sand, i.e. 56 cubic centimetres

of dry sand will absorb and hold between the grains 44 cubic centimetres of water. Coarse grades of sand have a lower capacity for
retaining water after the tide has fallen. There may also be a more
direct effect on the animals. Admixture of larger particles or pebbles
may make it difficult for some animals to burrow while worms which
live by swallowing sand are unable to do so if the particles exceed a
relatively small size.

Finally there is the effect of organic matter within the sand.
Especially in sheltered regions (and here we begin to merge into conditions more characteristic of muddy shores) there collects much
organic detritus, usually derived from the decomposition of seaweeds
growing on adjacent rocks. This detritus forms valuable food for many
animals and within limits adds to the productivity of the sandy shore,
but where it occurs in large amounts it tends to prevent free circulation
of water through the finely graded sand which accumulates in the
same regions. As a result supplies of oxygen in the lower levels cannot
be maintained and the population of certain bacteria which live in
the absence of oxygen increases. These bacteria are responsible for
the blackening of these layers owing to the formation by them of iron
sulphide. Such conditions, however, are not necessarily harmful to
burrowing animals if the blackening is not too great and they can
obtain oxygen from the water above.

Sand-dwelling animals must then be able in the first place to
burrow so as to survive the rush of waters over the beach and to escape
the effects of desiccation, lowered salinity and extremes of temperature.
But they must also be able to obtain food. They do this, broadly
speaking, in one of four ways. While buried in the sand they may, by
various means, draw in water from above when the tide is in and
from this collect suspended matter, largely plankton. These are
suspension feeders, like the barnacles, mussels and tube-worms of rocky
shores. Others collect food from the organic matter that collects on
the surface of the sand and consists in part of detritus of plant and
animal origin and in part of bottom-living diatoms and other minute
plants. Such animals are known as deposit feeders. Animals which
feed in these two ways can only do so when the tide is in.

A third group comprises animals that have in some measure freed
themselves from dependence on the water above and obtain food by
swallowing the sand and digesting the contained organic debris,
essentially as does an earthworm on land. Great quantities of sand

have to be passed through the gut because the amount of contained food is small. Some at least of such animals can feed when the tide is out so long as the sand remains saturated with water, and they may also respire under such conditions. Finally there are the inevitable carnivores which prey on the other sand burrowers. They are the most active of the inhabitants of sandy shores because they have to search out their prey and cannot remain quiescent, drawing in water or deposits or swallowing sand like the other animals.

Members of various of the major groups of the animal kingdom have become specialised for life in sand and they jointly provide what are possibly the best examples of adaptation to the same mode of life by animals of widely diverse form. The bulk of these sand dwellers consist of segmented worms, crustaceans and molluscs, with some highly specialised echinoderms and a few fishes. The procedure now will be to describe the commoner types of these animals and deal later with general matters such as distribution and zonation.

A tube-worm, the sand-mason, *Lanice conchilega*, is often very common on the lower half of sandy shores, especially where the beach has some protection and there are adjacent rocks with weed which provides a source of edible detritus. The worm constructs a flexible tube of sand grains, which projects for about one inch above the surface and is fringed with branched threads (Pl. XXIXa, p. 224). These projecting tubes are easily seen when the tide is out but there is no sign of the worm, which retreats deep into the tube when exposed or endangered. Although the animal may reach a length of up to ten inches the tube is considerably longer and only deep and fortunate digging will secure an intact specimen ; even then the body is so delicate that it ruptures unless carefully handled. It is a handsome worm, usually red with bright-scarlet gills, and in form not unlike *Cirratulus* with many long head tentacles as well as clusters of branched gills on the anterior segments of the body. When the tide is in and no danger threatens, the tentacles and gills are protruded for feeding and respiration. The animal is a deposit feeder and collects detritus from the surface of the sand with the tentacles, which also pick up sand grains for incorporation into the tube. The basis of this is a sticky mucus produced by pairs of what are known as gland shields on the structural under-side of the body. The sand grains are mixed with the mucus and added to the tube by means of the two-lobed lower lip. This can also repair the tube if the upper end is torn, while the animal can even construct a new tube if it is removed from the old one.

The body is devoid of the large lateral projections used in locomotion by the errant worms, such as the nereids, which would impede the necessarily rapid withdrawal into the tube, but it retains bristles of two types. The first seventeen segments of the body carry tufts of bristles which provide the necessary purchase for the animal to move up and down within the tube, while almost all segments possess great numbers of excessively minute hooks. By gripping the sides of the tube these enable the worm to make almost instantaneous retreat within its protection.

Another tube-worm found in sand, or in sandy mud, is of particular interest because it carries the tube about with it. *Pectinaria belgica* —it has no common name—constructs a firm tube of closely adherent sand grains, about an inch and a half long and wider at one end than the other. The grains of sand are carefully selected for size and fitted together with such precision that the tube is almost as smooth externally as it is internally. Although fragile and easily crushed, it is firmly built and may long persist after the death of the animal. This worm is commonest in the sub-littoral zone, but it may be found in sandy pools while the empty tubes, washed up after storms, are not uncommon objects on some shores.

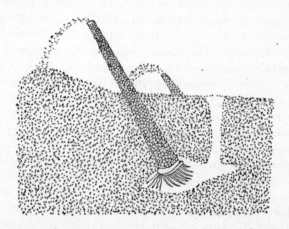

FIG. 65—Tube-worm, *Pectinaria*, burrowing and ejecting sand ; also shown are the horizontal burrow and the vertical shaft down which food falls from the surface deposits. Natural size. (Modified from Watson, *Proceedings and Transactions Liverpool Biological Society*, Vol. XLII, p. 25.)

The short, rather stout body is entirely protected within the tube with the head occupying the wider end. The body bears on each side a row of stout, golden-yellow bristles which are used in burrowing. The animal works its way obliquely into the sand head downward with the narrow end of the tube projecting for a short distance into the water above (Fig. 65, p. 218). Once buried it establishes further contact with the surface by way of a funnel-like opening through the sand. It then proceeds to feed largely on material from near the surface, and therefore rich in organic detritus, which descends this funnel and comes within range of the feeding tentacles. The water needed for respiration is drawn in through the narrow end of the tube, the opening of which can be plugged by the leaf-like hinder end of the body if there is danger of sand entering and blocking it. This mode of feeding demands some motility because the available food in any one place is soon exhausted, especially in the sub-littoral zone. The more superficial deposits on which the sand-mason feeds are continually being renewed by tidal movements so that immobile life in a fixed tube is no handicap.

The commonest of all sand-worms is the lob- or lug-worm, *Arenicola marina* (Pl. 36a, p. 177), which is responsible for the worm casts which are so conspicuous on many sandy beaches (Pl. XXIXb, p. 224). This worm is somewhat intermediate both in structure and habits between the errant bristle-worms such as *Nereis* and the tube-worms. Although it does not form a tube, it lives always under the sand and strengthens the walls of its burrow with mucus so that it really occupies a tube within the sand. It feeds like an earthworm by swallowing great quantities of the material into which it burrows and then digesting the small proportion of this which is edible. For this reason the lug-worm is commonest where there is some admixture of organic matter in the sand and it extends into regions where sand merges into mud.

The animal is relatively large, up to eight or nine inches long and as thick as a pencil. It is a very nutritious object to fishes and other predatory animals and for this reason is much sought after by fishermen as bait. The body is soft and easily damaged, when the red blood with which it is well supplied flows out. The blood is confined in a well-developed circulatory system, but the body cavity, between the body wall and the gut, is continuous and filled with coelomic fluid. This can be pushed forward or backward by appropriate contractions of the muscles of the body wall and in this way the body is kept firm and capable of movement within the burrow. The surface of the body

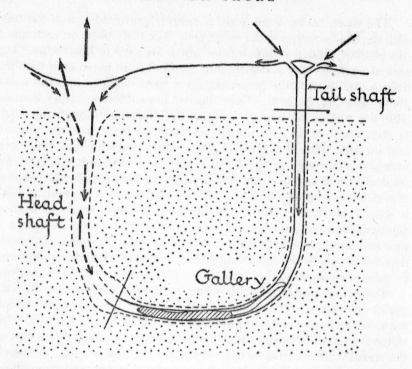

FIG. 66—Diagrammatic section of burrow of lug-worm, *Arenicola marina*, with worm in position, yellow sand shown white, dark under-layers of sand stippled, plain arrows denote direction of water currents created by worm, broken arrows movement of yellow surface sand into and down head shaft. (After Wells, *Journal Marine Biological Association*, Vol. XXVI, p. 170.)

is smooth, with much reduced bristles, but the middle region carries pairs of feathery gills. The hind region is thinner than the rest of the body and carries neither gills nor bristles. When burrowing, waves of contraction begin at the hinder end of the gill region and extend forward. These cause the animal to elongate and so move forward, because after the wave of contraction passes the body dilates and the protruded bristles grip the side of the burrow, preventing any backward movement. The head bears none of the sensory tentacles of the errant worms or of the more elaborate feeding tentacles of the tube-worms. These would be an encumbrance to a burrowing worm. But there is a large proboscis which can be extruded through the mouth and serves both to swallow sand and also, owing to its roughened surface, to grip the wall of the burrow.

The presence of *Arenicola* is quickly revealed after the tide falls by the appearance of the coiled castings and, close to these, of shallow depressions on the surface of the sand (Pl. XXIXb, p. 224). The castings consist of material ejected from the anus at the hind end, the depressions indicate the approximate position of the head end. The nature of the burrow and the habits of the animal within it have long been a subject of controversy. The problem has recently been reinvestigated in detail by Mr. G. P. Wells and the following account is based on his findings. The form of the burrow varies, especially in regions of shallow sand and mud where there are stones or shell fragments, but in the most suitable conditions of rippled sand with water retained in the depressions it is U-shaped as shown in Fig. 66 (p. 220). It can be conveniently divided into a gallery, tail shaft and head shaft. The gallery is the longest region and is L-shaped, part vertical and part horizontal, and here the worm normally lives. The walls are consolidated with mucus and marked by the fine hooked bristles with which the worm grips them when moving. It can move forward or backward in this region and when it moves back the hind region extends upward into the tail shaft—which never shows bristle marks, indicating that the middle or gill region of the worm does not enter. This backward movement occurs when the worm defecates and the castings are forced above the surface through one or more openings. The head shaft consists of a column of soft sand, yellow in contrast to the greyer colour of the deeper layers where there is some bacterial decomposition, and it may be as much as three times as wide as the gallery. It widens still further above into a cone underlying the surface depression.

The worm feeds by swallowing material at the base of the head shaft and this continual withdrawal of sand causes the depression above. But more than gravity goes to the formation of this region. There is evidence that the worm initially softens the sand above the end of the gallery both by pushing forward with a kind of piston action and also by extending the head region forward and then drawing this back again with the margins of the anterior segments extended like so many rounded flanges. Once the sand has been so worked it can be kept loose and open by a current of water which is driven through the burrow from the hind end forward by waves of contraction which continually pass along the body. This current is essential for respiration and the sand in the head shaft must be loosened sufficiently for the water to pass upward through it or else the worm would die for lack of oxygen. But once adequately loosened the upward stream of water keeps it in that condition. Meanwhile the worm continues to swallow sand and draw the loosened material down the shaft. The yellow colour of this sand is due in part to its origin from the surface layers and also possibly to some oxidation of the discoloured deeper sand by the flow of water through it.

Lug-worms do not move about to any extent and can certainly remain in the same position for weeks on end. The flowing tide fills in the depressions and adds new sand for them to swallow. The castings are usually of yellow sand but occasionally are black, indicating that the worms may swallow sand from lower levels despite its content of bacterial decomposition. The major danger to which the animals may be exposed is lack of oxygen when the sand is left bare and the burrow only partly filled with stagnant water. The rich content of red blood assists here because it is able to store oxygen for use over such a period while in addition the hind region may be coiled tightly, pushed upward above the surface of the water, and then drawn back with a trapped bubble of air which is brought in contact with the gills, which thus act temporarily as lungs.

The animals never voluntarily leave their burrows, even for reproduction. There is said to be a " genital crisis " lasting for two days during a period of rough weather about the middle or end of the first fortnight in October and then all the worms liberate their genital products into the water above so that fertilisation of the eggs is ensured. Like so many common and for that reason somewhat disregarded animals, the lug-worm is most highly adapted in structure, habits of life and mode of reproduction, for existence within its particular environment.

The worms remaining to be mentioned are carnivorous. There are a number of errant worms that move actively about within sand, the most conspicuous of which are probably *Nephthys hombergi*, the white worm or white cat, and species of *Glycera* (Fig. 43f, p. 150). Both are possessed of jaws borne on a relatively enormous introvert which may assist in burrowing as well as in seizing the prey. It is easily seen because when these worms are exposed this region of the gut is repeatedly extended and withdrawn. *Nephthys*, which is usually not more than about four inches long, may be recognised by the mother-of-pearl sheen on the body and the flattened upper- and under-surfaces. *Glycera* is yellowish in colour and often considerably larger with the rounded body characteristically tapering at both ends. It has the habit of reacting to disturbance by twisting the body into tight coils.

Among crustaceans the commonest of the sand-dwelling forms are undoubtedly minute copepods. These are not strictly speaking burrowers. They do not displace the sand, as a burrower must do, but are small enough to crawl over the surface of the grains of sand in the film of water carried upon these. They are incapable of the darting movements of the planktonic copepods and their much modified and elongated bodies are incessantly wriggling so that it is impossible to make out details of structure until they become moribund or die. Together with other microscopic animals, such as protozoans and thread-worms, they form what is known as the interstitial fauna. There are also great numbers of somewhat larger crustaceans which are true burrowers but which are yet too small to be adequately observed by the naked eye and must for that reason be passed over.

Possibly owing to the lateral flattening of the body, which may aid in burrowing, amphipods are much more common in sand than the isopods with which they are associated on rocky shores. The commonest of all is the sand-hopper, *Talitrus saltator* (Fig. 9, p. 28), which may occur in immense numbers ; " not millions, but cartloads " was the comment of one observer. It burrows in sand under weed and other debris along the strand line and is closely related to *Orchestia gammarellus* already encountered under decaying weed on the upper limits of rocky shores.

These hoppers are aptly named. They may jump for distances of several feet by a sudden straightening of the normally bent body, and if the weed on which they are feeding is disturbed their jumping bodies rise like a cloud above it. When food is scarce they may be seen jumping about in search of it. We may quote with approval the description, if not the explanation, encountered, surprisingly enough, in Paley's *Natural Theology*. " Walking by the sea-side, in a calm evening, upon a sandy shore, with an ebbing tide, I have frequently remarked the appearance of a dark cloud, or rather very thick mist, hanging over the edge of the water, to the height perhaps of half a yard, and of a breadth of two or three yards, stretching along the coast as far as the eye could reach, and always retiring with the water. When this cloud came to be examined, it proved to be nothing else than so much space filled with young *Shrimps*, in the act of bounding into the air from the shallow margin of the water, or from the wet sand. If any motion of a mute animal could express delight, it was this ; if they had meant to make signs of their happiness, they could not have done it more intelligibly. Suppose then, what I have no doubt of, each individual of this number to be in a state of positive enjoyment, what a sum, collectively, of gratification and pleasure we have here before our view."

The sand-hopper is nocturnal and burrows every morning in the dry sand high on the shore to emerge in the evening, when it may jump on the shore in an ecstasy one must attribute, despite the Archdeacon, to hunger rather than to joy. It feeds, like so many amphipods, indiscriminately on animal and plant matter. Four broods of up to seventeen young are produced annually between May and August and the female remains constantly in the burrow while carrying eggs or young in her brood pouch. The first brood reaches maturity by August or September but does not breed until the following year, when the young of the later broods also mature. Then in the autumn and winter of the second year all die so that the span of life is between a year and eighteen months.

Although they may range for some distance down the shore in search of food, these hoppers barely qualify as shore-animals. They usually feed on the jetsam above high tide level in company with a varied assemblage of truly terrestrial animals such as beetles, flesh-flies and small oligochaete worms. Both *Talitrus* and *Orchestia* contain species that are purely terrestrial while *T. saltator* is said to die of drowning if submerged for long periods although it does appear to need the presence of salt in the sand. These animals may fairly be grouped with the sea-slater and the rough periwinkle as creatures which have very nearly crossed the threshold of the shore on to dry land.

The amphipods of wet sand are numerous and occupy well-defined intertidal zones. There are three principal genera, *Bathyporeia*, *Urothoë* and *Haustorius*. All are capable of both swimming and burrowing but show interesting differences. *Bathyporeia* swims rapidly with the body upright and burrows with the aid of various of the appendages. *Urothoë* swims with somewhat less efficiency either erect or upside down ; and, as it burrows, uses the backward-flowing current created for this purpose by the swimming appendages of the abdomen to drive

PLATE XXIX

D. P. Wilson

a. Tubes of the Sand-mason, *Lanice conchilega*, exposed by the tide (×1)

D. P. Wilson

b. Casts and depressions caused by Lug-worms, *Arenicola marina*, on a shore of muddy sand

PLATE XXX

D. P. Wilson

a. Masked Crab, *Corystes cassivelaunus*, on sand, showing the upwardly extending antennae which form a breathing tube when the crab is buried. Photographed under water (×⅔)

D. P. Wilson

b. Burrowing Starfish, *Astropecten irregularis*. Photographed under water on the surface of sand (×¾)

out the sand. It continues, as it were, to swim right into the sand
although some of the legs assist it to burrow. The third type, *Haustorius,*
appears most specialised in structure and habit. The body is not so
flattened and the animal always swims upside down as shown in
Fig. 67, although it rights itself before starting to burrow. None

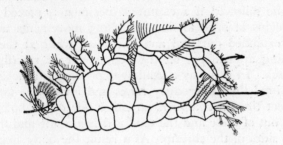

FIG. 67—Sand-burrowing amphipod, *Haustorius arenarius,* in swimming position,
arrows showing the main currents produced. Seven times natural size. (From
Dennell, *Journal Linnean Society, Zoology,* Vol. XXXVIII, p. 363.)

of these animals forms a tube in the sand and all probably feed
on minute particles of organic matter in the water held within the
sand or adherent to its grains. While finding both protection and
food within the sand they may leave this periodically for reproduction.
Bathyporeia is most frequently found swimming and always at night.
It appears in greatest numbers at about fortnightly intervals, controlled
possibly by the height of the tides. The eggs take some fifteen days to
develop and it is reasonable to assume that fertilisation at one period
of freedom in the sea is followed in the next by hatching and liberation
of the young.

The best-known of all sand-dwelling crustaceans is certainly the
common shrimp, *Crangon vulgaris* (Fig. 11, p. 29). On suitable shores
it is extremely common and easily caught at low tide by pushing a
net with a basal cross-piece of wood along the surface of the sand in
shallow water. Shrimps are closely allied to prawns and the differences
in structure are largely associated with differences in habit. Prawns
have already been described in some detail and it is worth while seeing
how this type of animal has become adapted for life on a soft substratum of sand. The body of the shrimp is more flattened from above
to below so that the appendages of the middle and hinder regions are

more widely separated. The sharp spine or rostrum, which projects conspicuously forward from between the eyes in the prawns is absent. Only the first of the five pairs of legs carries pincers but these limbs are very stout and capable of dealing with much larger objects than the two pairs of finer pincers borne by the prawns.

During the daytime shrimps burrow in the sand. The process can easily be followed if a captured specimen is placed in a sandy pool. It begins at once to make shuffling movements with the legs which are extended outward and backward while at the same time the swimming appendages on the abdomen beat rapidly and drive the sand back. In this way a shallow depression is quickly excavated and the animal then alternately raises and lowers the lateral shields which protect the gill chambers on either side of the body. Water is thus forced out of these and the sand is driven aside and then upward around the sides of the shrimp. As a result the body sinks gradually down. Finally, by a kind of breast-stroke action, the long feelers push sand around and over the back and the body is completely covered. These antennae are also usually withdrawn below the surface later, but two small branches continue to project and by their means food is detected.

Even when exposed, shrimps are not easy to see, because the colour of the body tones closely with that of the background—and although this may range from clean yellow sand to much darker mixtures of sand and mud, the animal can change colour to match this through the agency of its pigment cells. Free movement is normally confined to the hours of darkness when the animals often move slowly along the surface on the last two pairs of legs, the other three being tucked away along each side of the mouth. The third, and longest, pair project in front and are often employed as tactile organs, tapping in appraisal any hard object encountered. Swimming is also possible but is indulged in less frequently and with less efficiency than by the prawns.

Little that is edible comes amiss to shrimps and the diet varies throughout the year. At times the stomach can be seen through the translucent body packed with ingested green plants, at others the food may be exclusively animal and include smaller crustaceans, molluscs, eggs and young stages of fish, and even relatively large worms. Shrimps have been seen devouring ragworms larger than themselves. And always there are organic remains of all kinds as a basic food supply for these omnivorous creatures.

Few shore animals are more hardy than shrimps. They cannot, it is true, survive more than brief exposure to the air, but, like all sand-dwellers, this is an experience they are unlikely to suffer. But within or upon the surface of the sand they can withstand temperatures ranging from freezing point up to 30° C. and can flourish in low salinities so that they are often common in estuaries where we shall again encounter them in a later chapter.

In their mode of growth and reproduction shrimps closely resemble prawns, but, except in estuarine waters, they are less sensitive to the cold of winter and may spawn in that period, although the developing eggs are then carried for up to thirteen weeks, or three times as long as they take to develop in the summer. Owing to the shape and habits of the animal, the female lies on her side during copulation and the eggs are attached to the base of the last two pairs of walking legs as well as to the swimming appendages.

Sandy beaches in the tropics are the home of a fascinating array of burrowing crabs, many of them semi-terrestrial in habit. The only counterpart on our temperate shores is the masked crab, *Corystes cassivelaunus* (Pl. XXXa, p. 225), and even this is really an inhabitant of the sub-littoral zone and only rarely found at extreme low water of spring tides. But it is an admirable example of adaptation to life in sand and justifies a short description. It lives exclusively in clean sand, unlike the burrowing " prawns " found in muddy sand and described in the next chapter. *Corystes* does not shuffle about sideways like most crabs but when placed on sand proceeds to sit upright with the oval carapace vertical. The furrows on the back of this give it a certain resemblance to a human face and are responsible for the common name. A pair of antennae, as long as the body, project upward and each bears a double row of stiff hairs which are directed inward so that when the two antennae are brought together these hairs interdigitate and form a tube which communicates at its base with the gill chambers. Such a tube forms an admirable channel for drawing water through a coarse medium like sand but would be useless in mud.

The crab wastes no time in starting to burrow, by means of the four hinder pairs of legs each of which terminates in a long claw. These dig into the sand and pull the animal down, the sand being thrown up on the structural under-side of the animal by the scooping action of these legs aided by the long first pair which carry pincers. The latter are much larger in the male so that the sexes can be distinguished at a glance. Gradually the animal disappears from view but burrowing continues until only the tips of the antennae project above the surface. Water is then drawn down the tube between the closely applied antennae and passes over the gills on either side of the body from the front of the chamber backward. In all surface-living crabs and allied crustaceans this respiratory current flows in the opposite direction, as can easily be seen if colouring matter is added to the water in which they are kept. There is a sound reason for this because the anterior openings of the gill chambers lie one on each side of the mouth and a backward-flowing current would draw in fragments of food and so foul the gills. Even in the masked crab the current passes in this direction when the animal is not buried, the change after burrowing being associated with the use of the antennae to form a tube for conveying clean water to the buried animal. Like other crustacean burrowers, *Corystes* probably burrows mainly during the daytime to escape from enemies such as skates and other fish and emerges at night to forage for food.

Of all marine animals, the bivalve molluscs are the most perfectly adapted for life within soft substrata of sand or mud. Their fundamental characters, including the lateral flattening of the body and foot, the enclosure within a hinged, bivalved shell, and the great development of the gills within the enlarged gill chamber so created, all unite to this end. With the evolution of the bivalve form the molluscs were able to exploit the rich possibilities of life beneath the protective surface of sand or mud while drawing in food from surface deposits or from the suspended plankton. While it is true that some took to attached life secured by byssus threads or cement to a hard surface, the great majority of bivalves remain burrowers and the further development of this habit, with suitable modifications of shell and foot, has led to the successful boring into rock and timber already described.

The bivalves of sandy shores may be divided into two major groups according to the nature of their food, which involves some difference in the mode of collection. There are plankton, or suspension, feeders and there are deposit feeders. The best example of the first group is the common or edible cockle, *Cardium edule* (Fig. 68, p. 229). In suitable areas this exists in almost astronomical numbers; the population of a single cockle bed of some 320 acres on a sandy shore in South Wales where the animals live literally " cheek by jowl " has been estimated on the basis of careful sampling at round about 462 millions !

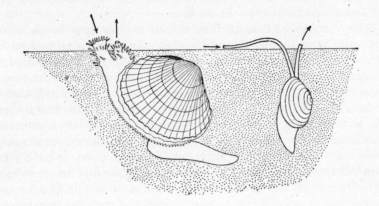

FIG. 68—Sand-burrowing bivalves : (left) cockle, *Cardium edule*, a suspension feeder and (right) *Tellina tenuis*, a deposit feeder, showing positions in sand, shape of foot and inward passage of water and food and outward passage of water through the siphons. Natural size.

The shell of the cockle is well known ; it is more than usually globular for a bivalve and is marked with conspicuous radial ribs which possibly assist the animal to grip the sand. The foot is a large and powerful organ as it has reason to be if it is to move so cumbersome an animal through and over sand. By its aid the cockle burrows, but only deep enough to cover the shell, as shown in Fig. 68, and also " leaps " on the surface. This latter process is not quite so spectacular as it sounds ; the animal protrudes the rounded, fleshy foot which is bent in the middle and then, with the tip pressed against the sand, suddenly straightens the foot like a released spring, and the cockle usually rolls over several times, although it may sometimes skip in the air if the jerk is sufficiently powerful.

Cockles spend most of their time quietly buried in the sand with the two short siphons projecting above the surface when the tide is in. Apart from their length, the siphons are essentially similar to those of the boring bivalves and are united. Through the lower one water enters the gill chamber, to be strained through the meshwork of the gills, and leaves by the upper one. The food thus consists entirely of fine particles, largely microscopic plants of the plankton, which are suspended in the water and the immense population of rich cockle beds is an indication of the great abundance of such suspended food in the water that flows over the sands.

Cockles appear. to flourish best in mid to low shore levels where the tide runs most rapidly to and fro ; although they occur in greatest density on regions of clean sand they are by no means confined to such areas. They are often found in regions of mixed mud and sand and sometimes extend well within estuaries on a muddy substratum and where the salinity is low. Under these conditions the shell is more lightly built, possesses fewer ribs and is more asymmetrical, although why lowered salinity should have these effects remains an unsolved problem. Cockles can move freely on the surface, as we have seen, and on occasion whole beds of them may apparently migrate—this is probably due to water movements. The story is told of a clergy- man near Morecambe Bay who complained that half his parishioners had left him. The local industry was cockle-gathering and the animals had moved from the shores bounding his parish to others across the Bay, whither his parishioners had followed them.

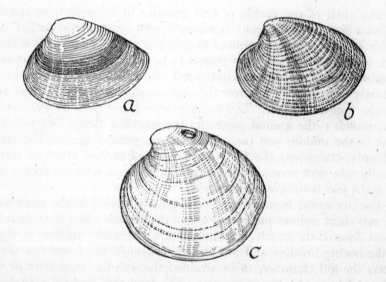

FIG. 69—Suspension-feeding bivalves : *a. Spisula subtruncata* ; *b. Venus gallina* ; *c. Dosinia lupinus* with bore hole made by the carnivorous snail, *Natica*, near the hinge. Left shell valves, natural size.

Other sand-dwelling bivalves which feed in the same manner include species of *Spisula* (*Mactra*), *Venus*, *Dosinia* (Fig. 69, p. 230) and of the razor-shells, *Ensis* (*Solen*). The first three are typical bivalves with shells more flattened than those of cockles and so capable of somewhat deeper and more rapid movement in sand. When exposed their immediate reaction is to regain security within the sand ; they make no ungainly progress over its surface. The siphons are short so that the animals must lie near to the surface when feeding, although they descend lower when the tide is out. The genera, if not the individual species, are easily distinguished. The shell of *Spisula* is triangular and smooth (Fig. 69a). That of *Venus* is also somewhat triangular but more rounded and with well-marked concentric rings ; the commonest species between tide marks is *V. gallina* (Fig. 69b), which is up to an inch long and usually has three reddish rays running from the hinge to the margin of the shell valves. *Dosinia lupinus* (Fig. 69c) is somewhat larger with a smoothly rounded shell possessing a pronounced hollow, or lunule, in front of the hinge.

The razor-shells are the most highly specialised of sand-burrowing bivalves and possess a unique power of rapid downward movement to escape danger from enemies or from desiccation. The elongated shell is a common object on the shore and cannot be mistaken for that of any other bivalve. Two species may be dug from clean sand near low water level. In the larger, *Ensis siliqua*, the shell is almost straight and cut away almost at right angles at each end. It may be as much as seven inches long and over an inch broad. *E. ensis* has a smaller and somewhat curved shell with one end more rounded than the other. Examination of these animals, or of empty shell valves found cast up on the shore, shows that the hinge lies close to one end—the anterior. The great elongation of the shell is thus due to prolongation of the hinder half. At this end the short siphons project while the enormous foot, which occupies at least half the space between the shell valves when it is withdrawn, is pushed out anteriorly and not on the structural under-surface, opposite to the hinge, as it is in typical bivalves such as cockles. Actually in all bivalves the foot is directed forward after extrusion and the animal is pulled along by it, directed either diagonally down or horizontally beneath the surface, with the anterior end in front. In the razor-shells movement is always vertical with the foot, extended in the long axis of the shell, leading the way as it burrows down. The straight, smooth shell offers the minimum of resistance to passage through the sand so that the movement is exceptionally rapid.

When the tide is in, the animals approach near enough to the surface for the short siphons to project above this, but when the water leaves the sand they retreat deeper, although their presence may be indicated by shallow depressions from which sudden jets of water and sand may be forced up by the animal. To be caught they must be approached with caution because razor-shells are highly sensitive to vibrations and at once retreat still deeper. A sudden deep dig with a spade or fork may bring up an intact specimen, more often a broken portion, but most frequently, except to the expert, be fruitless. A less exhausting mode of capture is to place a handful of salt over the hole. As this dissolves the increased salinity irritates the animal below and it may come to the surface and project the hinder end of the shell sufficiently for this to be seized and, with a sudden jerk, pulled out intact. Any hesitation will give the foot time to take grip of the sand below and it may either succeed in pulling the shell down or else the animal may be literally torn in two between the opposing pull of hand above and foot below.

This remarkably tenacious grip is due to the manner in which the foot operates. As shown in Fig. 70 (p. 233), the tip of this is pointed as it extends down and so it runs easily through the sand. When fully extended, that is when the internal blood spaces have been distended to their utmost by blood forced into them from the rest of the body, the tip of the foot swells out into a bulbous disc. This grips the sand like a mushroom anchor while the retractor muscles that run from the foot to the shell contract and pull the animal down. At the same time the blood flows back from the interior of the foot into the body. During their downward movement the shell valves are drawn closely together by the adductor muscles, but when the foot is protruded in the next stage of downward movement the valves separate and so hold the animal firmly in its temporary position. But for this action the shell would tend to move up as the foot descends.

Although razor-shells probably never leave the sand of their own volition, storms may expose them. They are then quite capable of making their way back into the sand, a process which can be readily observed by anyone who has obtained an intact specimen. It is illustrated in Fig. 70. The foot is protruded and immediately turned downward around the margin of the hinge and so into the sand where, after extending to its full length, the tip swells out in the manner described above. As the shell valves are drawn together, owing to the simultaneous contractions of the muscles of the foot and shell, water is forced out around the sides of the foot and this assists in washing away the sand and facilitates the entrance of the shell. The first movement is usually sufficient to carry the animal some distance below the surface; it is immediately repeated and in a few minutes even the largest razor-shell will have disappeared under the surface of the sand.

Upward movement, essential if the animal is to regain contact with the sea after it has burrowed deeply to escape danger or exposure, is performed with no less efficiency. The foot can push the animal up as well as pull it down. In this pushing process the tip of the foot only is initially distended with blood while the retractor muscles remain relaxed. With the end of the foot thus anchored, blood is forced into its upper regions which elongate while the shell valves are drawn together. Upward movement is the inevitable result and the process can be repeated if necessary by withdrawal of the foot and further

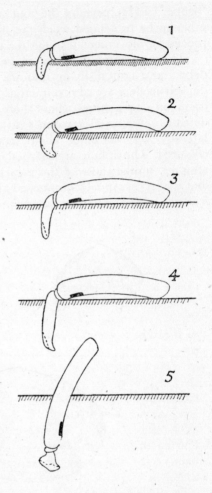

FIG. 70—Burrowing movement of the razor-shell, *Ensis*. (From Russell, *The Behaviour of Animals*, after Fraenkel.)

anchoring at higher levels. These up-and-down movements are beautifully executed and are one of the distinctive features of these highly specialised burrowers. In relation to them it should be noted that neither end of the shell valves ever closes; they are said technically to "gape." This permits the foot at one end and the siphons at the other to be protruded when the shell is otherwise closed in the actual process of movement through the sand.

The remaining bivalves to be discussed all feed on the organic debris and bottom-living diatoms on the surface of the sand. These deposit feeders are also numerous in mud but species of three genera occur on sandy shores where there is some admixture of finely divided detritus. *Tellina, Donax* and *Gari (Psammobia)* all possess smooth and laterally very flattened shells, many of which are very beautifully coloured. The siphons are characteristically long and separated from one another, unlike those of the suspension feeders (Fig. 68, p. 229). The foot is also very much compressed although relatively large. All of these features in their anatomy are related to their particular mode of life.

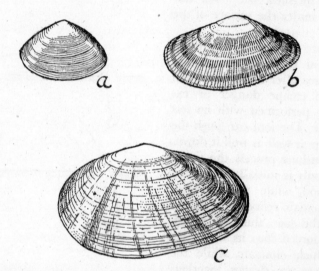

FIG. 71—Deposit-feeding bivalves : *a. Tellina tenuis* ; *b. Donax vittatus* ; *c. Gari depressa.* Left shell valves, natural size.

Tellina tenuis (Fig. 71a) is the commonest of these bivalves. It has an oval shell, delicate and translucent and often beautifully tinted with pink, yellow or orange. It is a true intertidal species, occurring in small numbers about the level of high water of neap tides, increasing in density down to the level of low water of spring tides, and finally disappearing at a depth of about three fathoms. In the zone of maximum abundance counts of over one thousand per square metre are common and some over eight thousand have been recorded. They grow slowly to lengths of about half an inch in three years—somewhat more quickly in the upper zones : in these zones they are less numerous than lower on the shore, where the competition for food must be most intense. A second species, *T. fabula*, a little smaller and more angular in shape with characteristic diagonal striations on the right valve only, overlaps with *T. tenuis* around low water level but extends to greater depths ; it is essentially a sub-littoral species. Other species live exclusively on the sea bottom and only their empty shells will be found on the shore.

Donax vittatus (Fig. 71b, p. 234) occurs more locally, although sometimes in considerable numbers, and always near low water level. It is known as the purple toothed-shell and is one of the most beautiful of British bivalves. The shell is roughly oblong and about an inch long. It is much stouter than *Tellina* and very highly polished. The shell margins are finely toothed and the inner surface is usually coloured violet although externally it is most often white, though yellow, brown and violet shells are not infrequent. Species of *Gari* are most common off-shore but some may be found between tide marks. The shell is again much compressed but more angular and less smooth than in the preceding genus. The one most likely to be found living on the shore, especially in the south, is *G. depressa* (Fig. 71c), a little over an inch long and known as the setting-sun-shell on account of the rays of pink that run from the hinge outward to the margin of the shell.

All of these bivalves are admirably fitted for rapid movement through sand. The shell is compressed, to such an extent in some, especially in *Tellina*, that it seems hardly possible for an adequate body to be housed between the closely applied valves, so that the minimum of resistance is presented to movement brought about by the compressed foot, which seems to flow like a thin film downward and forward from between the opened shell valves. Such movements are necessary both for deep retreat under the sand when the tide is out and for horizontal passage to new feeding grounds. During feeding the siphons are extended, the upper one, through which the water current is expelled, projecting for a short distance above the sand, while the other, which is longer and very mobile, curls over and explores the surface of the sand as shown in Fig. 68 (p. 229). The force of the inflowing current created by the ciliated gills is revealed by the stream of particles which pours along this siphon and into the gill chamber. This siphon is constantly moving and extends in all directions and for varying distances until all available food is sucked in. Tidal movements will to large measure replenish this but these animals do need to move about, unlike the suspension feeders which are often stationary unless dislodged by the force of the sea.

The large quantities of fine bottom material taken in by the inflowing siphon are not all swallowed. There is some sorting in the gill chamber by the gills themselves and by the paired flaps or palps that guard the small mouth-opening, but the great bulk of what is taken in is ejected through the siphon by which it entered. The adductor muscles make periodic contractions which reduce the volume of the gill cavity and force out this excess material. Such minor muscular contractions for cleansing the gill chamber occur in all bivalves, although less frequently in the suspension feeders where excess material collects much more slowly. A special region of the adductor muscles is concerned with these sudden contractions which are quite distinct from the slow and long continued contractions, which occur when the shell valves are closed for protection. In the swimming scallops and in *Lima hians* this " quick " portion of the muscle is enlarged and provides the power for the continuous flapping of the valves during swimming. In the scallops it is even striated like the muscles attached to the skeleton in vertebrates ; the other portion, the " catch " muscle, is composed of smooth fibres like the muscles surrounding the gut and arteries in vertebrates.

Univalve molluscs are not common on our sandy shores. Here again tropical shores abound in sand-burrowing snails but we have only one that is common, the carnivore *Natica alderi*, which preys on small bivalves, especially *Tellina* and *Venus*. Like all sand-burrowing snails it has an immense foot with which it ploughs its way slowly along just under the surface. Encountering a bivalve it grips this with the foot and proceeds to make a neat round hole through the shell before eating out the contents by means of the inserted proboscis (Fig. 72, p. 237). In the dog-whelk, it may be recollected, the shell of the prey is bored through mechanically but *Natica* employs chemical means through the agency of a disk-shaped gland on the under-side of the proboscis. Bivalve shells may often be found perforated with a

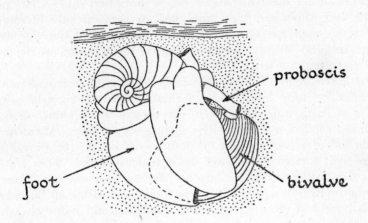

FIG. 72—Carnivorous sand-burrowing snail, *Natica*, attacking a bivalve which it is holding with the lobes of its large foot while boring through the shell by means of the proboscis. (From Hesse, Allee and Schmidt, *Ecological Animal Geography*, after Schiemenz.)

neat hole about a millimetre in diameter and tapering from the outside inwards, as shown in Fig. 69c (p. 230). These holes are certain indications of attack by *Natica*. The animal will only be found by digging, but is easily identified by the rounded and very smooth shell. It lays characteristic spawn in the form of spirals of jelly strengthened and thickly encrusted with sand grains.

Sand-burrowing echinoderms include a starfish, a sea-urchin and a form of sea-cucumber. *Astropecten irregularis* (Pl. XXXb, p. 225), the starfish, is frequently common on sandy bottoms offshore and may occasionally be dug at low tide on the beach. It is of particular interest because it shows how the typically crawling starfishes can become adapted for burrowing. The body is very flattened and the five arms have a bordering of large plates and of projecting spines. The tube-feet are pointed, not sucker-like as in the common *Asterias*, and are of little use for locomotion on a hard surface. The animal cannot climb up the side of a rock. The tube-feet are adapted for digging and when placed on sand this starfish quickly disappears from view, sinking straight down. It is a voracious carnivore like all starfishes and feeds on molluscs, crustaceans, worms and other echinoderms which it encounters under the sand, and even on moribund fishes. The prey is swallowed whole and the empty shell or skeleton later disgorged. There is no open passage to the surface of the sand but the water current essential for respiration is created by vibration of the long spines that fringe the arms.

The heart-urchin, *Echinocardium cordatum*, is often very common in sand near low tide level. It is a most interesting creature which displays striking differences from the regular sea-urchins, such as *Echinus*, which live on rocks. The shell is very fragile. Water-worn tests devoid of spines are not uncommon objects on the strand line and are light enough to be blown by a moderate wind. They are not spherical as in the regular urchins but slightly longer in one diameter and roughly heart-shaped. Owing to their habit of life all burrowing urchins have lost the original radial symmetry and move always in

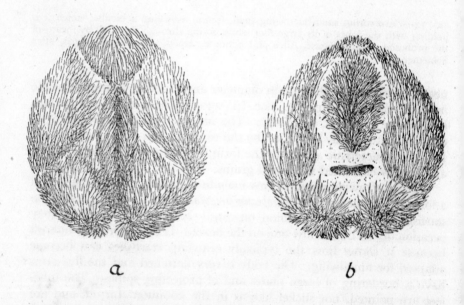

a   *b*

FIG. 73—Sand-burrowing urchin, *Echinocardium cordatum*; *a.* aboral (upper) view; *b.* oral (under) view. The former shows the five petal-like grooves between the long backwardly-directed spines, the latter the transverse mouth opening with the spatulate digging spines behind this. Natural size. (From Borradaile, Eastham, Potts and Saunders, *The Invertebrata*.)

FIG. 74—Sand-burrowing urchin, *Echinocardium cordatum*, within sand. The tube to the surface is consolidated by a sticky substance. Through it water is drawn for respiration while special tube-feet (one of which is shown) extend on to the surface of the sand above for feeding. (From Hesse, Allee and Schmidt, *Ecological Animal Geography*, after v. Uexküll.)

one direction. We can, therefore, speak of a front and a hind end, which we cannot say for *Echinus*. The holes through which the tube-feet emerge in life are obvious on a bared test. They are largely confined to five wide diverging rows, like the petals of a flower, on the upper surface. One of these lies within a groove which indicates the front end of the animal. A few other tube-feet project through holes around and behind the mouth on the under-surface. This has maintained its original central position, but the anus has shifted for some distance back on the upper surface. In an intact specimen (Fig. 73, p. 238), such as may be obtained by digging, the test is covered with delicate golden-yellow spines, backwardly directed in the main and varying in size and function. The largest and most interesting are a group behind the mouth which are broad and spatulate. Spines on either side of the anterior groove bend inwards and protect the channel.

These structural details are necessary for an understanding of the mode of burrowing and of the activities of the animal within the sand. The spines are the burrowing organs, especially the spatulate ones on

the under-surface which actively dig, but the other spines assist and
all move freely. The urchin burrows straight down to a depth of six
to eight inches but retains contact with the surface by way of a channel
as shown in Fig. 74 (p. 239). The walls of the burrow and of this
channel are consolidated with a sticky mucus so that the sand does
not fall in and the animal thus lies free within the cavity it has made.
It is a deposit feeder, collecting food from the surface by means of the
very extensile tube-feet belonging to the front row of its upper surface.
These extend up the channel and on to the surface of the sand where
they pick up particles of edible matter and pass them, by way of the
well-protected groove, to the mouth below. There the tube-feet from
that region take over and place the food in the small mouth. There
are no teeth and no massive Aristotle's lantern, which would be useless
to an animal which feeds on small particles. The new position of the
anus, well clear of the top of the animal where food is brought down
by the collecting tube-feet, is associated with the altered mode of life.
Water currents for respiration are created by cilia and certain of the
tube-feet act as gills.

Feeding as they do, it seems improbable that heart-urchins remain
for long in the same position. Although never actually observed, it has
been suggested that they come to the surface at night and, after moving
for some distance, make a new burrow. On the other hand, they have
been observed in aquaria to change position without coming to the
surface, although how they re-establish contact with it remains un-
determined. There is still a good deal to learn about the habits of this
most interesting animal. In suitable areas it is very abundant and
would appear to be truly gregarious, a habit certainly of value during
spawning since it increases the chances of fertilisation of the freely
liberated eggs. Reproduction begins during mid-summer in the second
year of life. After initial planktonic life, settlement takes place in the
sub-littoral zone and the young animals migrate upward above low
tide level during the first year.

The tough-bodied sea-cucumbers of rocky shores are represented
among sand-burrowers by the worm-like Synaptidae such as *Leptosynapta inhaerens*, allied to *Labidoplax digitata* (Pl. XXXIb) which will
be mentioned in the next chapter. *L. inhaerens* is a delicate creature
about three inches long with a pink translucent body and lives in much
the same regions as *Echinocardium*. Tube-feet are confined to a ring
round the mouth and concerned solely with feeding. The anchor

PLATE·XXXI

D. P. Wilson

a. Cuttlefish, *Sepia officinalis*. Photographed under water ($\times \frac{1}{2}$)

D. P. Wilson

b. Burrowing Sea-cucumber, *Labidoplax digitata*, with the commensal animals which live in its burrow, the Scale Worm, *Harmothoë lunulata*, and the Bivalve, *Mysella bidentata*
Photographed under water ($\times 1$)

PLATE XXXII

D. P. Wilson

a. Collecting Peacock Worms, *Sabella pavonina*, from the muddy shores of the Salcombe estuary at extreme low water of spring tides

A. J. Lloyd

b. Shrimp or hose nets on the intertidal mud flats at Stolford, Bridgwater Bay, Bristol Channel

spicules embedded in the thick body wall of other sea-cucumbers here lie with the flukes projecting from the surface. They are at once detected when the body is touched and are a sure indication of its nature. In life they serve to grip the sand while the tentacles push sand and organic debris into the mouth. The highly muscular body wall constricts and severs readily when handled and, as the head end can probably regenerate what is lost, this habit is equivalent, on a somewhat wider scale, to the autotomy of starfishes and of crabs.

The sand-living fishes, the sand-goby, young flatfishes, lesser weaver and sand-eels, have all been mentioned in the chapter dealing with shore fishes. All are adapted in form, habit and coloration for life on or within sand and play their part largely as carnivores in the economy of life on sandy shores.

Zonation exists on sandy shores as it does on those of rock but is concealed beneath an unbroken surface of sand. It can be revealed only by systematic digging at various levels, taking similar areas in each case and carefully sieving the sand ; a laborious business which takes time and enthusiasm. The results, moreover, are not always too easy to interpret in terms of the relation of the animals to environmental conditions. Where the shore slopes steeply the higher regions from which water drains away at low tide will be largely barren, but a gently sloping beach which retains ample water at all levels will be populated throughout. Desiccation plays no part in zonation which would seem to be determined largely by the amount of time that a particular animal needs to be covered in order to obtain adequate food. Cockles are largest near low water level but *Tellina tenuis* is larger at higher levels. This may be explained by the longer time taken by a suspension feeder to obtain adequate food from the plankton; *Tellina* certainly draws in the deposits on which it feeds with remarkable rapidity and should be able to obtain adequate food in a correspondingly shorter time. The relatively sharp distinction between the zones occupied by the sub-littoral *Tellina fabula* and the shore-dwelling *T. tenuis* is not easy to explain. There is a possibility, but it is untested, that the pressure of water may have some effect.

Sand-living animals attain their final position on the beach in various ways. The bivalves settle there from the plankton, larval lugworms develop on the bottom and then burrow direct, the amphipods emerge in the form of small adults from the maternal brood pouch. In all cases future life is confined to a zone the upper limit of which is imposed probably by the effects of reduced immersion and the lower limit by more uncertain factors, which may include competition with other species and increased water pressure. Other animals, such as the burrowing urchin, move upward when half-grown after settlement in the sub-littoral zone. They go so far and no further, possibly again restrained by limitation of the period available for feeding. All sand-living echinoderms and fishes keep low on the shore and represent a sub-littoral population which has achieved minor penetration up the shore. Molluscs and most worms are widely distributed over the shore but in the main spend larval life in the plankton. The sand-living crustaceans, mainly amphipods, which never leave the shore show the widest distribution of all as shown in Fig. 75, which indicates the

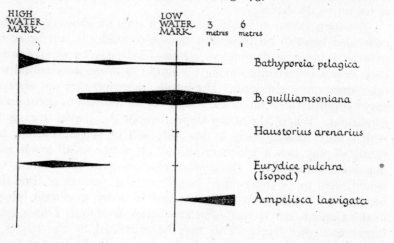

FIG. 75—Vertical distribution of four species of amphipod crustaceans and one isopod on the sandy shore at Kames Bay, I. of Cumbrae. (Modified after Elmhirst, *Proceedings Royal Society, Edinburgh*, Vol. LI, p. 169.)

distribution of five species of amphipods and one isopod on a sandy shore near the marine station at Millport.

The apparently lifeless expanse of exposed sand has been shown to harbour unsuspected hosts of animals, diverse in origin but with common powers of burrowing. The abundance, under favourable conditions, of cockles, spisulas, tellinas and lug-worms reveals the wealth of food provided by suspended plankton or by finely divided organic deposits.

CHAPTER 16

MUDDY SHORES

"And often, standing on the shore at low tide, has one longed to walk on and in under the waves . . . and a solemn beauty and meaning has invested the old Greek fable of Glaucus the fisherman, how he ate of the herb which gave his fish strength to leap back into their native element, was seized on the spot with a strange longing to follow them under the waves, and became for ever a companion of the fair semi-human forms with which the Hellenic poets peopled their sunny bays and firths, feeding his ' silent flocks ' far below on the green Zostera beds, or basking with them on the sunny ledges in the summer noon, or wandering in the still bays on sultry nights amid the choir of Amphitrite and her sea-nymphs."

CHARLES KINGSLEY : *Glaucus ; or, The Wonders of the Shore*. 1855

WHEREVER the force of the sea is abated finer material is deposited and sand begins to give place to mud. At first this is still firm with a large admixture of sand or gravel and often with interspersed boulders, but eventually it merges into a soft slime over which progress becomes increasingly difficult and finally impossible. No sharp line can be drawn dividing the faunas of sand and mud. There is a transitional area inhabited by a variety of animals that live in a stiffish mixture of the two and only a relatively limited community of highly specialised animals can exist in the clogging environment of a true mud bank. Muddy shores are hardly ideal for a pleasant seaside ramble, their attractions are only for the fisherman and the keen naturalist, but no account of the sea shore would be complete without some description of their more typical inhabitants.

The shallow sheltered waters at the entrance to a muddy creek or small estuary are the home of the eel-grass, *Zostera*, the one flowering plant that is truly marine. It cannot withstand the greater exposure of sandy beaches and needs, moreover, a relatively stable bottom of muddy sand in which the long roots can ramify. The smaller *Z. nana* is widely exposed at low tides ; but only the upper limits of the wider area covered with the larger *Z. marina* are exposed. This occupies the same zone as the tangle-weeds off rocky shores, and like them, can best be observed from a small boat and more easily because the surface of these sheltered waters is often unruffled.

Extensive beds of the larger eel-grass were formerly widespread on both sides of the Atlantic. Then, beginning about 1931, they were almost completely wiped out by a disease which seems to have originated on the American side. Only in the Mediterranean and on the Pacific coast of North America were they unaffected. Various possible causes of the disease, protozoans, fungi and bacteria, were found by investigators in the dying plants but certainty eluded them. Many marine animals, and also ducks, geese and swans, which used to live or feed upon the *Zostera*, either died out or moved to other feeding grounds. A considerable industry based on the grass, which had a variety of commercial uses, being used, after drying, for filling mattresses and upholstering furniture, was destroyed. Natural populations are prone to such large-scale destruction and especially in the sea where the causal agent may be carried widely in ocean currents. Within more recent years the commercial sponges around the Bahamas and Florida have suffered similar destruction although in this case fungal infection is the certain cause.

There is little evidence as yet of any return of the eel-grass, but in time the former beds may gradually be re-established. The smaller and much less important *Z. nana* was not affected. Any description of such a bed is based on knowledge of the past and inspired by hope for the future. The still waters between the long-extending leaves were the haunt of pipe-fishes and sometimes also of the common cuttlefish, *Sepia officinalis* (Pl. XXXIa, p. 240). The most magnificently illustrated book on marine life ever published, Méheut's *Étude de la Mer*, which all interested in marine life should see and aspire to possess, contains among its many coloured plates one of these animals, their striped bodies appearing through the leaves of eel-grass like tigers in a jungle. A number of small snails made their home exclusively on these leaves to which were attached the small trumpet-shaped lucernarians— coelenterates belonging to the same group as the large jellyfish but in which the entire life is spent in the hydroid phase. *Zostera* beds

---

PLATE 37

a. PEACOCK-WORMS, *Sabella pavonia*, with the large SIMPLE SEA-SQUIRT, *Phallusia mammillata*

b. FAN-WORM, *Bispira volutacornis*

PLATE 37

Photographs by D. P. Wilson

PLATE 38

Photographs by D. P. Wilson

were also the home of the sea-hare, *Aplysia* (Pl. 34a, p. 169), and of a number of similar soft-bodied tectibranch snails. It was a quiet environment rich in rather unusual life and beauty and we can but hope for its eventual return.

Shore conditions vary widely where mud is deposited and description must be confined to broad generalities. But it is possible to distinguish between muddy inlets where the water is fully saline throughout and estuaries up which the sea is increasingly diluted with fresh water and in which the mud content may be enormously increased by material carried down by the river and deposited where this mingles with the sea. Fine particles of mud carried in suspension in river water coagulate into larger masses and fall to the bottom when the water becomes salty. The problems of life in the low salinity of estuaries are discussed in the next chapter.

We may suitably look first at shores of muddy sand and at others with scattered stones or small boulders such as are often encountered near the mouths of inlets before sand is completely replaced by fine mud. An area of muddy sand is often the site of a rich bed of the gaper, *Mya arenaria* (Fig. 76, p. 246), the soft-shelled clam of America, where it is equally abundant and where the natural beds are often protected owing to its use as food. This is among the largest of British bivalves, a good specimen being six inches long and about three inches broad. The shell is darkly coloured owing to the thick outer horny layer which is continued over the massive siphons. When, after some exertion, an animal is dug out, the siphons may have retracted, but their contracted mass can be seen between the hind ends of the shell valves, where there is a permanent gape due to the relatively enormous mass of the siphons. There is no gape anteriorly, as there is in the razor-shell, because the foot does not need to make frequent sudden movements. An allied species, *M. truncata*, which is somewhat smaller, occurs in stiff mud. Yet another gaper, *Lutraria lutraria*, occurs in

PLATE 38

a. SPONGE, ~~*Hymeniacidon sanguinea*, growing on a stone upon a muddy estuarine shore~~

b. STONE from lowest tidal levels at Salstone, Salcombe, showing Compound Sea-squirts, *Morchellium argus* and *Amaroucium nordmanni* (with tuft of red weed, *Antithamnionella*, growing in middle), also grey and pink specimens of the Slug, *Lamellaria perspicua*

a. SPONGE, HALICHONDRIA BOWERBANKI, SIMPLE SEA-SQUIRT, ASCIDIELLA ASPERSA AND COLONIAL SEA-SQUIRT, MORCHELLIUM ARGUS, ALL CHARACTERISTIC OF VERY SHELTERED WATERS, EXPOSED AT EXTREME LOW WATER OF SPRING TIDES AT THE SALSTONE, SALCOMBE ESTUARY.

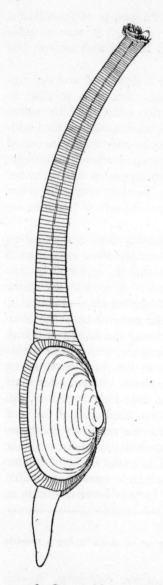

clean sand. It is superficially very like *M. arenaria*, but belongs to a different family of bivalves.

In early life *M. arenaria* has a large foot and can move about freely, and also makes temporary attachment with byssus threads (Fig. 77, p. 247). But with increasing size it takes to a passive life beneath the surface, where it dwells vertically embedded in the stiffish substratum, slowly descending as it grows and its siphons elongate. These may eventually attain a length of over a foot. Their openings, fringed with tentacles, lie flush with the surface at the bottom of a shallow depression which is the sole indication of the presence of the animal a foot or more below. Digging out *Mya* is hard work but it has not the exasperating qualities of a similar search for razor-shells. The massive body of the clam cannot slide downward like the smooth straight shell of *Ensis* and the animal makes no such attempt. With increasing size the foot gets relatively smaller and it is only called into play in the most unlikely event of the animal being forced out of its deep burrow. Fully grown specimens kept in captivity on a suitable substratum in an aquarium will burrow anew, but it is a most laborious process taking many days to complete. It is most unlikely that the animal would be left unattacked in nature long enough to enable it to rebury itself. Once deeply embedded in a suitably stiff substratum it must be immune to danger from storms and spend a safe but motionless existence collecting plankton and other suspended particles from the water above.

FIG. 76—Gaper, *Mya arenaria*, with siphons and foot fully extended. One-half natural size. (After Meyer and Möbius.)

FIG. 77—Young *Mya arenaria*, 4 mm. long, showing relatively large foot, shorter siphons and byssus threads present for attachment during early life. (From Kellogg, *Shell-Fish Industries*.)

Other molluscs are common in this region of mingled sand and mud. Cockles persist, although never in the same numbers as on clean sandy beaches. *Tellina tenuis* is replaced by another type of deposit-feeding bivalve, *Macoma (Tellina) balthica*. This is somewhat longer and possesses a stouter and somewhat more rotund shell. It congregates, sometimes in immense numbers, in the more muddy patches from the surface of which the indrawing siphon collects with continuous activity immense quantities of the superficial deposits. Food is more abundant here than on the sandy shores where *Tellina* lives and change of position is less frequently necessary ; this may account for the somewhat more globular shell, which may be associated with decreased mobility. Where there are scattered stones mussels will be found attached to them with the ever common periwinkles crawling over them, while shore-crabs will shelter around their bases or in hollows on the surface of the shore while the tide is out. Such shores are the natural home of small whelks and the smaller carnivorous snails, *Nassarius (Nassa) reticulatus* and the sting-winkle *Ocenebra (Murex) erinacea*, all of which have a long siphon which is held clear of the mud.

Regions of muddy sand, often where there is *Zostera*, may be inhabited by the daisy-anemone, *Cereus* (Pl. 40, p. 253), which here attaches itself to stones well below the surface. These are also the most likely places to find specimens of the true burrowing-anemone, *Peachia hastata*, which has a rounded base and occupies a burrow sometimes nearly a foot deep. It has twelve long tentacles and these, with the central disc, are of a yellowish brown with beautiful markings. When feeding it projects for a little distance above the mouth of the burrow but may retire within this for protection. The larvae pass into the sea where they become for a time parasitic upon jellyfish or the medusae of various hydroids. The name *Peachia*, bestowed by Gosse, commemorates Charles Peach, a self-taught Cornish naturalist who was a member of the coastguard service in the first half of the last century. On a daily wage of four shillings he supported and largely educated a family of nine children and yet found time to add many new species to the British marine fauna.

A varied population of worms live on and upon this muddy sand. Many have already been encountered, chiefly on sandy shores, such as lug-worms, the sand-mason, the brown paddle-worm, *Phyllodoce maculata*, and the carnivorous " white cat," *Nephthys hombergi*. All diminish in numbers as the soft mud comes to predominate over the firmer sand. One of the ragworms, *Nereis diversicolor*, recognised by the conspicuous red line down its back, appears where the salinity is lowered. It is a typical estuarine species and will be further considered in the next chapter.

Digging at low water of spring tides on shores of muddy sand may expose one of the most interesting of all marine worms, the sea-mouse, *Aphrodite aculeata* (Pl. 39b, p. 252). It is essentially an inhabitant of the sub-littoral and is most often found on the shore after it has been thrown up there during stormy weather. It is quite the most massive of British species, not at all worm-like in appearance but with a broad flat body up to six inches long and two inches wide. It is a scale-worm, allied to the common and much smaller *Polynoë* of rocky shores, and bears fifteen pairs of large scales, or elytra, on the back. But these are entirely obscured in the intact animal by a dense covering of matted hairs, unusually long and thin bristles, which form a felting over the back. Laterally the animal bears other bristles, some stout and short and used in locomotion and others, probably protective, which shine with a brilliant iridescence, with golds and reds and other colours, and give to this obscure inhabitant of the sea bottom a quite surprising beauty. The sea-mouse lives beneath the surface with only the hind end exposed. By appropriate body movements water is drawn in along the under-side and then ascends in streams between the laterally projecting parapodia and so into the respiratory space between the back of the animal and the over-arching scales. The latter are then drawn down and the water expelled by way of the upper side of the tail. The thick felting above and the lateral iridescent bristles protect this respiratory current from contamination or blockage by the surrounding muddy sand (Fig. 78). The internal arrangements are no less striking. There is an immense muscular region behind the mouth-opening where the carrion which the animal encounters and swallows in its slow movements is ground up. The fragments are then squeezed back into a series of fine tubes running off laterally from the hinder portion of the gut, in which the final processes of digestion and absorption are conducted.

The more completely protected muddy shores of fully saline creeks

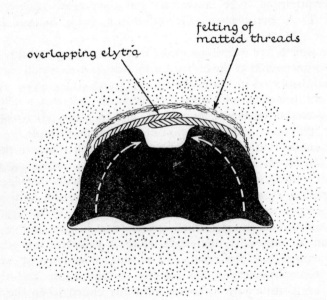

FIG. 78—Sea-mouse, *Aphrodite aculeata*, diagrammatic cross-section of animal under mud (indicated by stippling). Water is drawn in below the projecting tail region and passes forward along the channels indicated, then upward between the laterally projecting parapodia as shown by the white broken arrows into the space between the elytra (scales) and the back where respiration occurs. The water is then passed backward. (Modified after van Dam, *Journal Experimental Biology*, Vol. XVII, p. 1.)

and drowned valleys, such as many of the shallower sea lochs on the west of Scotland and the much studied " estuary " at Salcombe in south Devon (Pl. XXXIIa, p. 241), are often particularly rich collecting grounds. There may be veritable forests of the projecting tube-worms, notably the peacock-worm, *Sabella pavonina* (Pl. 37a, p. 244), which constructs a firm muddy tube up to a foot long. From this tube emerges a wide ring of brightly coloured tentacles which, richly ciliated, collect food from the water well clear of the bottom mud. *Branchiomma vesiculosum*, with a smaller crown of tentacles, utilises fragments of coarse sand and gravel in the construction of its tube, and a third species, *Myxicola infundibulum*, is easily distinguished by the very thick gelatinous tube without any incorporated matter. The tentacles of this last species are webbed for more than half of their length, so forming a particularly massive and complex crown. Excellent figures, in line and colour, and descriptions of these fan-worms and their crowns are given in Professor T. A. Stephenson's delightful little book, *Seashore Life and Pattern*.

These worms live where stones are frequent amongst muddy gravel; and stones uncovered at the ebb of spring tides possess a rich encrusting fauna well displayed in Plates 38b (p. 245) and 40 (p. 253). Another type of crumb-of-bread-sponge, *Halichondria bowerbanki* (Pl. 25b, p. 144), is characteristically found here and so is the red *Hymeniacidon sanguinea* (Pl. 38a). The population of colonial and simple sea-squirts is exceptionally rich. The most abundant of each type are probably respectively *Morchellium argus* and *Ascidiella aspersa*, shown together in Plate 25b. But the most impressive is undoubtedly the largest of British species, *Phallusia mammillata*, which lives in the same regions as the peacock-worm, with which it appears in Plate 37a. This massive simple sea-squirt, which grows to a height of five or six inches, has an unusually thick milk-white or yellowish test of a firm, almost cartilaginous consistency. The surface is covered with the rounded elevations responsible for the specific name and the two openings are conspicuous although naturally closed when the animal is found uncovered by the tide. On cutting through the test, the internal organs can be removed intact and examination made of the large branchial basket through which water is strained for food collection, together with the opaque mass of the other viscera. Both this animal and the smaller *Ascidiella aspersa* may here be picked up unattached, a sure indication of untroubled waters, because both are

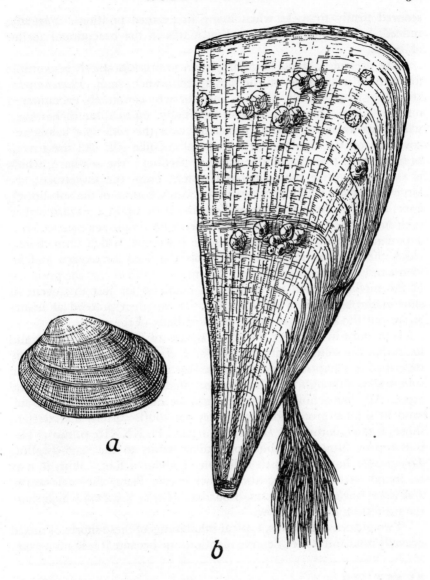

FIG. 79—Bivalve molluscs: *a.* carpet-shell, *Paphia* (*Tapes*) *pullastra* ; *b.* fan-shell, *Pinna fragilis*, showing long byssus threads. Both natural size. (Latter from Jeffreys, *British Conchology*.)

secured firmly to rocks when living in exposed positions. *Phallusia*, indeed, is only found between tide marks in the exceptional shelter of inlets.

The presence of these large simple sea-squirts is probably responsible for annual incursions of the large tectibranch snail, *Pleurobranchus* (*Oscanius*) *membranaceus*, which is not otherwise commonly encountered on the shore. This has a flattish oval body, up to three inches long, with a brownish-red mantle projecting over the wide foot below and covering, on the right side, the large plume-like gill. All these tecti-branchs have their particular mode of feeding ; the sea-hare, which is also common in these sheltered waters, crops the sea-lettuce; the large *Scaphander*, which burrows in the sandy bottom of the sub-littoral zone, swallows small bivalves and crushes them up in a gizzard armed with limy plates ; while *Pleurobranchus* feeds on simple sea-squirts. It is a patient animal. It crawls on to the sea-squirt, which immediately closes the two openings. But inevitably the need for oxygen and for food causes eventual reopening. The moment this occurs the proboscis of the mollusc, which it has held in readiness for just this event, is shot in through one of the openings. It can then proceed at leisure to eat out the soft tissues of the enclosed body of the sea-squirt.

It is only in the very sheltered waters of fully saline creeks and lochs that the solitary British feather-star, *Antedon bifida* (*rosacea*) briefly described in Chapter 3 and shown in Fig. 16 (p. 37), appears between tide marks, although it is abundant on mud bottoms in the sub-littoral zone. Another echinoderm which may be dug from mixed sand and mud in similar protected areas along our south-western and western shores is the holothurian, *Labidoplax digitata* (Pl. XXXIb, p. 240). This is a red or brownish animal of similar habits to the sand-dwelling *Leptosynapta* but growing to a length of about a foot. With it may be found its commensals, also shown in the Plate, the scale-worm, *Harmothoë lunulata*, and the small bivalve, *Mysella bidentata*, which share the protection of the burrow.

Two genera of bivalves, typical inhabitants of these shores of mixed gravelly mud and stones, deserve mention, one because it is so numerous,

PLATE 39
a. BURROWING "PRAWN," *Callianassa subterranea*
b. SEA-MOUSE, *Aphrodite aculeata*, seen from above and showing the lateral spines and the protective felting over the back

PLATE 39

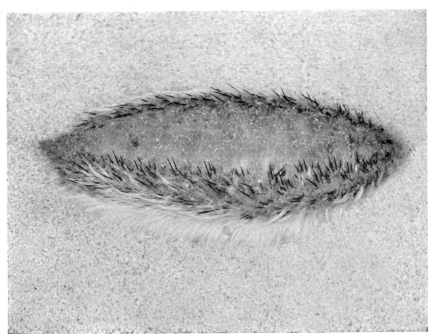

Photographs by D. P. Wilson

PLATE 40

the other because it is rare but of particular interest. The carpet-shells, *Paphia (Tapes) pullastra* (Fig. 79, p. 251) and *P. decussata*, are amongst our commonest bivalves wherever there is a bottom of muddy gravel, and for this reason are not uncommon in such a substratum under and between boulders on a rocky shore. But they may be dug in large numbers where there is a wide expanse of suitable shore. They possess a stout oblong shell, rounded in front and truncated behind, a little under two inches long. The species are a little difficult to separate but the shell of *P. decussata* has coarser ridges. They are suspension feeders, like cockles, and rather take the place of these animals on shores of gravelly mud, into which they may burrow somewhat deeper owing to the greater length of the fused siphons.

The fan-shell, *Pinna fragilis* (Fig. 79), has the distinction of being the largest British bivalve. Specimens up to fifteen inches long and eight inches wide have been found, but it is rare between tide marks although common, albeit difficult to obtain, in deeper water. It is much flattened laterally with the shape of a partly opened fan (or that of a small ham, the French call it "jambonneau") and lives vertically buried in muddy gravel with the pointed anterior end undermost and the free edges of the shell at the wide posterior end projecting a little above the surface. It is distantly related to the common mussel and has the same type of shell, although in a much exaggerated form, and an even better developed byssus, the threads of which are attached to stones deep below the surface. *Pinna* bears a name bestowed by Aristotle. It is very common in the Mediterranean, where its byssus threads, which are exceptionally long and tough and resemble silk, have long been used commercially. They were woven into fabrics and fashioned into garments worn by Roman emperors. The industry has never completely died out, persisting particularly at Taranto where there are immense natural beds of these bivalves, but the gloves, caps, stockings and other smaller garments made from the byssus threads are now objects of curiosity rather than real economic use. Attempts were made in Victorian days to develop a similar industry in this country and Gwyn Jeffreys, writing in 1863, reports that " at our last International Exhibition a Cornish muff made of this material might have been seen by those who were disposed to venture into an obscure gallery in search of the few objects of natural history for which any space was allotted."

Among the comparative rarities of sandy mud are the burrowing " prawns " (really anomurans allied to the hermits and the squat lobsters) *Callianassa subterranea* (Pl. 39a, p. 252), *Upogebia stellata* and *U. deltaura*. The first is entirely confined to the south coast, the other two occur also on the east ; all three may be collected in the Salcombe estuary. They are essentially sub-littoral animals, present in great numbers at moderate depths but difficult to obtain from there, with just sufficient vertical range to bring them on to the lowest levels of well protected shores where they may be dug out at low water of the bigger spring tides. Merely to view these animals, with the small carapace and large abdomen, soft and loosely put together, is to see at once that they are not adapted for the swift movement of the true prawns. The eyes are reduced and the hinder, " swimming " appendages are small and primarily concerned with creating a respiratory current through the burrow. *Upogebia* lives in pairs within deep and permanent burrows with several openings to the surface. There it finds the protection for which it has abandoned speed of movement and the harder integument of the surface-living prawns. Food consists of finely divided material which is strained from the water by the hairy mouth-parts in much the same manner as in the allied porcelain crabs. The burrows of *Upogebia stellata* are the home also of the rare bivalve, *Lepton squamosum*, a beautiful animal with a pure-white shell, and a fringe of tentacles.

The fauna of the banks of soft mud found beneath the turbid waters in the upper reaches of estuaries is often abundant but confined to the relatively few animals that can feed and respire in this excessively soft and finely divided substratum. There will also be problems of lowered salinity but those we must leave for the moment. An animal which can maintain itself in sand or sandy mud will flounder in the softer mud and the mechanisms of feeding and of respiration will become clogged with the fine particles. A few animals have adapted themselves with striking success to life in and upon this medium, most notably the deposit-feeding bivalve, *Scrobicularia plana*, the little snail, *Hydrobia ulvae*, and the amphipod, *Corophium volutator*, all of which burrow into it, with the ragworm, *Nereis diversicolor*, and the common shrimp which move actively upon or through it.

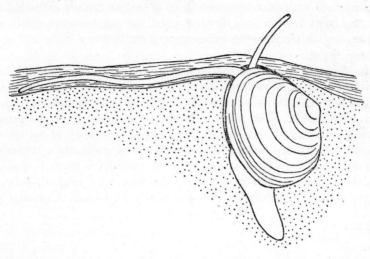

FIG. 80—Mud-living deposit-feeding bivalve, *Scrobicularia plana*, showing mode of feeding. The animal normally lives much deeper, up to six or eight inches below the surface. (From Hesse, Allee and Schmidt, *Ecological Animal Geography*, after Meyer and Möbius.)

Scrobicularia (Fig. 80) is the largest of the deposit-feeding bivalves. The oval shell grows to lengths of from one and a half to two inches and is extremely flat, more like that of the sand-dwelling *Tellina* than the other mud-dweller, *Macoma*. It lives between tide marks, and where the salinity is lowered by fresh water ; and always in finely graded mud in which there is usually a high content of organic matter —in such mud densities of over 100 per square metre have been found. Despite the flat shell and large foot which enable it to burrow deeply, it seldom moves about. This may well be due to the rich supplies of food in the surface mud which the mobile siphons suck in with untiring activity while the tidal flats are covered by the sea. The force of the inflowing current is such that the tip of the siphon seems almost to tear away the surface of the mud. When the tide is out, the position of the animal below is revealed by the characteristic star-shaped markings made by the intake siphon as it extends in all direc-tions in search of food. *Scrobicularia* is well worth collecting, especially if really large specimens are available, to see these siphons. When the animals are dug up these will be withdrawn but if left undisturbed in a jar of sea water the siphons will be at first cautiously extruded and then extended further and further until the jar becomes filled with pairs of mobile transparent tubes some 2 millimetres wide and up to six inches long.

Hydrobia ulvae is a small snail not more than a third of an inch high. Thomas Pennant, who first described it, said it was the " size of a grain of wheat." Individually insignificant, it is frequently impressive in its abundance. The surface of the mud may have a granular covering of these animals which can exist at a density of up to sixty thousand per square metre. The specific name indicates a preference for a diet of the sea-lettuce, *Ulva*, but the animal is more catholic in its tastes and must feed on surface vegetation and organic debris of all kinds. Like *Scrobicularia*, it is essentially an estuarine animal and

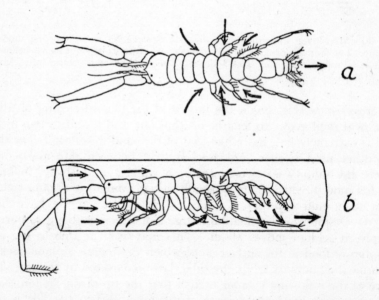

FIG. 81—Mud-burrowing amphipod, *Corophium volutator*, showing nature of water currents created : *a.* when crawling freely over the bottom ; *b.* when enclosed in a tube. About seven times natural size. (From Hart, *Journal Marine Biological Association*, Vol. XVI, p. 761.)

commonest on the higher levels of mud flats, though it extends beyond them on to the less frequently covered areas of salt marshes. Although it rises to the surface of the mud when this is uncovered, it burrows to depths of about a third of an inch when the tide is in. The upward movement at uncovering is probably due to inability to maintain contact with the surface when the soft mud settles as the water drains away from it.

Corophium volutator is another typical inhabitant of relatively soft mud banks provided there is no excessive blackening due to organic decay and some admixture of fresh water. Under such conditions this little amphipod, about a third of an inch long, may occur in immense numbers, inhabiting straight or U-shaped burrows up to five inches deep (Fig. 81, p. 256). The walls of the burrow are maintained by means of a sticky substance formed by glands in the second pair of legs. The span of life is short, probably less than a year because two generations appear to be produced annually, the young hatched in the spring and early summer becoming mature the same year and their offspring surviving the winter to breed in the following year. The young animals may make side passages from the parent burrow in much the same manner as do the young of the wood-boring gribble.

Like so many of the other amphipods mentioned, *Corophium* feeds on organic detritus. It emerges from the burrow to crawl over the surface of the mud and pick up fragments of edible matter with the first pair of legs while the appendages of the abdomen draw in a current of water laterally beneath the body for respiration (Fig. 81a). When the animal is within the burrow this water current passes along the length of the body, as shown in Fig. 81b, and suspended matter carried in with it is filtered off as it passes below the head and is then probably sucked into the mouth.

Shrimps appear as well able to thrive in mud as in sand, although they are not uninfluenced, as we shall see, by lowered salinity. On estuarine mud flats they may find rich feeding in the immense populations of *Nereis diversicolor* which crawls over the surface. Fish are often common enough in the muddy waters and one at least, the conger-eel, *Conger vulgaris*, often lives under stones among mud. There it may be exposed at the ebb of spring tides where it even provides sport to the local inhabitants. At Kilve, and neighbouring villages on the Somerset coast, congers or " glatts " are hunted with dogs and C. G. Harper in *The Somerset Coast* has given a lively description of this unique sport.

" There is not the smartness among the pursuers of the glatt which is the mark of the hunting-field in the chase of the fox or the deer, and renders a fox-hunt or a meet of staghounds so spectacular a sight. Smart clothes are not the proper equipment of the glatt-hunter, whose hunting chiefly consists in wading, ankle-deep, through the mud, heaving up great boulders, and mud-whacking after the wriggling, writhing congers, while the dogs rush frantically among the crowd, scraping holes in the mud and essaying the not very easy task of seizing the slippery fish." The results of a day's sport are congers averaging about four or five pounds in weight with occasional specimens up to twenty pounds.

CHAPTER 17

LIFE IN ESTUARIES

"Also this year on the tenth of April, 1607, there was a strange Fish caught in King Road and brought to the Back in a boat of Cardiff, the fish was called a Fryar being five foot in length and three foot in breadth having two Hands and two Feet and very 'Grisley and Wide Mouthed, he was Hall'd on a Hallier's Dray to Mr. Mayor's House."

Bristol City Council, *Minutes of the General Committee Book.* 1801

T H E presence of fresh water has an immediate effect upon the nature of the shore population. Even where a trickle of fresh water runs over a rocky shore its course is marked by a rivulet of green sea-weed that descends through the zone of brown fucoids. Similar green weeds form the only flora of pools at and above high tide level where fresh water predominates and in which dwell a characteristic population of brackish-water animals capable of living in water of low and often widely varying salinity. Where rivers enter the sea there is a region, sometimes of great extent, of mingled fresh and saline water which is encountered by the animals and plants of the shore when the bounding headlands are passed and the shores on the two sides, converging slowly, become the boundaries of the estuary before they merge higher up with the river banks.

An estuary, so far as we are here concerned, may be defined as the region of varying salinity between the mouth, where the water is fully saline, and the head where it is fully fresh. The physical conditions within estuaries are often very complicated, and, as the subject lies only on the fringe of the subject-matter of this book, it will be sufficient to speak of them in the most general terms before proceeding to describe the nature of the population much of which represents upward migrants from the shore.

The waters of the rivers are always moving seaward ; those of the sea rise and fall with the tides. If the river bed is sufficiently steep, such as that of a mountain torrent that runs directly into the sea, the force of the fresh water will be great enough to prevent the sea entering and the river will be fresh to its mouth ; estuarine conditions as we are here considering them will not exist. The other extreme is represented by the inlets and creeks mentioned in the last chapter where negligible amounts of fresh water enter and where there are only the up-and-down movements of the tides with no appreciable seaward flow of fresh water to affect salinity or to raise mud from the bottom. The typical estuary is one in which the river current, after long passage down slowly descending valleys and then over almost level plains, has insufficient force at the mouth to prevent the entry of the sea when the tide is rising. Then the seaward flow is stemmed, to be resumed with correspondingly greater speed when the tide ebbs. Such an estuary resembles the shore: there is a rise and fall of tidal waters, but the upper zone in mid-channel is fresh water and not land. Owing to the damming of the river waters, the effect of the tides may be experienced far above the estuaries in regions where the water is fresh. This is displayed most dramatically in the Severn where, at spring tides, the influence of the bore in the estuary below causes an appreciable rise in the river water, known locally as a quarrage, as high up as Worcester.

It is possible to divide estuaries into two main types depending on the manner in which sea water enters. This may push forward with the rising tide as a compact mass, mingling in front with the fresh water so that the salinity gradually falls towards the head of the estuary but does not at any point vary significantly from surface to bottom. This is the case, for example, in the Severn estuary. In other estuaries, such as that of the Tees, there is a vertical salinity gradient. When the tide begins to rise, the sea water flows in along the bottom while the fresh water continues to move seaward, although with diminishing speed, nearer the surface. Hence the salinity at the surface will be low but will rise with increasing depth. The difference between these two types of estuaries is shown in Fig. 82 (p. 261).

It follows from this that marine animals capable of life in this oscillating mass of brackish water within the confines of the estuary will be able to penetrate for a considerable distance up the bottom of an estuary with a vertical salinity gradient, but upward progress will soon cease along the shores which are washed by the fresher surface waters. When there is no such gradient and the saline waters push forward equally at all depths, then the salinity between tide marks will gradually decline up the estuary. While it is true that the salinity will rise and fall with the tides, low salinity will coincide with low tide, when the shores will in any case be largely uncovered and the population protected in the various ways already described. Of more significance

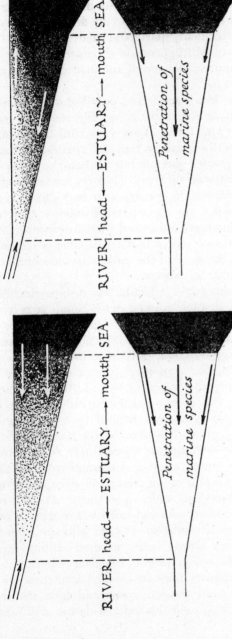

Estuary without vertical
salinity gradient

Estuary with vertical
salinity gradient

SEA ←mouth← ESTUARY ←head RIVER

SEA ←mouth← ESTUARY ←head RIVER

Penetration of
marine species

Penetration of
marine species

FIG. 82—Types of estuaries; sections (above) showing penetration of sea water at high tide and surface views (below) showing penetration of marine species along shores and in mid-channel. (With acknowledgments for assistance to Mr. R. Bassindale.)

will be the difference between neap tides and spring tides because the saline waters do not penetrate so far up the estuaries on the neaps. The distance to which any particular marine animal or plant can extend up an estuary will depend on the lowest salinity which it can withstand for any appreciable period and that will be largely controlled by the salinity at neaps.

So far we have been considering only the variability represented by the waxing and waning of the tides. But the river waters are not constant in amount and force. They vary with the rainfall and with evaporation. Generally speaking they are greatest in the winter, not entirely owing to greater rainfall but also to the lower evaporation during the cold months of the year. Then the sea water will be unable to push its way so far up the estuary as it does during a hot and dry summer. The extension up an estuary of animals that live passively on the shore will be largely influenced by the downward movement of fresh water in the winter ; those that are capable of moving to and fro will migrate nearer the head of the estuary in the summer and retreat nearer to the mouth in the winter.

Another important factor affecting life in estuaries is the amount of sediment carried in their waters. This comes largely, though not always entirely, from the rivers and tends to settle where fresh water mixes with the sea, forming the conspicuous mud banks so often characteristic of large estuaries. Much of this material is carried in suspension, more at spring than at neap tides and more in winter than in summer, rendering the water turbid. Little light can penetrate. Recordings made in the Tamar estuary have shown that the intensity of light at the surface is always reduced to 1 per cent within four metres and on occasion in less than half a metre. Life is not easy under such conditions. Plants are immediately affected by the reduction in light, while many animals are smothered, or their delicate mechanisms for respiration or for feeding clogged, with the fine sediment that swirls around and settles upon them. For this reason alone the rich fauna of muddy creeks and inlets where there is only a gentle rise and fall of the tides insufficient to raise sediment from the bottom, is absent in estuaries. Nevertheless lowered salinity represents the major difficulty in the colonisation of estuarine waters.

The fauna of estuaries may be divided into three groups. First these are stragglers from the sea below and from the rivers above. These are animals that can withstand a wide range of salinity. Since many shore animals have this power, it is not surprising to find a variety of them, such as the shore-crab and the common periwinkle—the eggs of the latter have been found capable of developing in water much below full salinity—extending for considerable distances up the shores of estuaries. Some fish have even greater powers, notably the flounder, which, although usually regarded as a marine animal, can pass through estuaries and live in fresh waters. And there are many occasional visitors into estuaries, such as the Fryar (monk or angler-fish, *Lophius piscatorius*) recorded in the extract at the head of this chapter, a six-foot specimen of which was caught high in the Severn estuary at Berkeley in 1939. Second there are those fishes which during their lifetime normally pass from the sea to the rivers or vice versa, such as the *anadromous* salmon, sea-trout and shad, which enter the rivers to spawn, and the *catadromous* eels which, after feeding and reaching maturity in fresh waters, pass down to the sea to spawn. Both kinds of fish make at least two journeys through estuarine waters in the course of their lives. But they are only transitory inhabitants of such waters and need not detain us.

Finally, and most important, there are the truly brackish-water animals which are found only in water of lowered and variable salinity. These are *euryhaline* animals, possessed of the functional capacity to maintain the concentration of salts in their blood and body fluids irrespective of changes in that of the water around them. Such powers are not possessed by purely marine animals ; when the sea water is diluted there is immediate inflow of water into their bodies and death ensues. Freshwater animals have this power ; most of them acquired it originally by initial adaptation to life in estuarine waters, but they have lost the power of resisting changes and so, like the higher animals on the shore and the terrestrial animals, cannot retrace their steps and pass back into estuaries or to the sea.

The matter can best be illustrated by reference to the habitat of different species of the common amphipod crustacean, *Gammarus*. A number of species of this genus are known which are restricted to different environments, so closely indeed that the presence of a particular species is a clear indication of the nature of the water. They are examples of biological indicators. It will be sufficient for us to discuss three species, the very common and exclusively freshwater *G. pulex*, the equally abundant marine *G. locusta* which is common on our shores, and the no less specialised *G. zaddachi* which is a

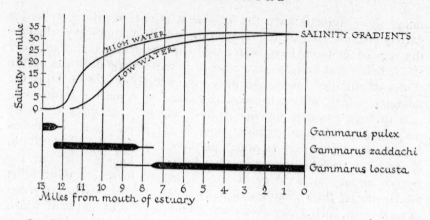

FIG. 83—Distribution of species of the amphipod, *Gammarus*, in the estuary of the R. Deben, Suffolk. For full explanation see text. (Modified from Serventy, *Internationale Revue der gesamten Hydrobiologie und Hydrographie*, Vol. 32, p. 286.)

purely brackish-water species. The distribution of these three species around the estuary of the river Deben in Suffolk, together with the prevailing salinities, is shown in Fig. 83. *G. pulex*, it will be seen, does not extend below the head of the estuary ; *G. locusta* penetrates for only a short distance within its mouth ; while *G. zaddachi* is restricted to the zone of brackish water of changeable salinity within the confines of the estuary. The story may be carried a stage further, because it has recently been shown by Mr. G. M. Spooner that *G. zaddachi* is divisible into subspecies. Slight structural differences between *G. z. salinus* and *G. z. zaddachi* are associated with functional differences which confine the latter to higher estuarine zones than the former. A third subspecies, *G. z. oceanicus*, is a northern form and occurs in Scotland, where it occupies the zone inhabited by *G. z. salinus* further south.

Interesting experiments have been carried out on different species of the common ragworm, *Nereis*, revealing that the species characteristic of brackish and estuarine waters, *N. diversicolor*, has much greater powers of controlling the concentration of the blood and body fluids than have the common species on the sea shore. Other brackish-water species, similarly equipped, include the small, almost transparent, prawn, *Palaemonetes varians*, which is often very common in pools along sandy stretches bounding estuaries. These stretches also harbour certain species of opossum-shrimps or mysids, such as *Praunus flexuosus*

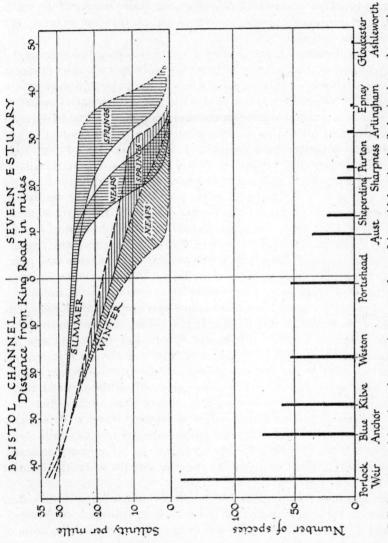

FIG. 84—Salinity ranges (above) and approximate numbers of intertidal animals (below) along the southern shores of the Bristol Channel and Severn Estuary. Full explanation in text. (Prepared from figures and data by Bassindale, *Journal of Ecology*, Vol. XXXI, p. 1. and *Proceedings Bristol Naturalists' Society* (4), Vol. IX, p. 386.)

(Fig. 10, p. 29). Estuarine mud flats, the home of *N. diversicolor*, are often covered with a granular mass consisting of immense numbers of the small brackish water-snail, *Hydrobia ulvae*, while species of the small burrowing amphipod, *Corophium*, mentioned in the last chapter, are also typically estuarine.

In a small estuary there is often a relatively abrupt change in the shore population from marine to brackish-water species. But in larger inlets there may be a great area of water outside the technical mouth of the estuary where fresh water dilutes the sea. Under such conditions there will be a gradual change in the shore population, due to decreasing salinity and often also increasing turbidity, before coming to the true estuary with its characteristic population. In this country such conditions are best displayed in the Bristol Channel and Severn estuary. Surveys of the waters and life in this area were made while I was at Bristol and it will be fitting to refer here to some of the results, in particular those of my friend and former colleague, Mr. R. Bassindale.

Much of what has to be related is recorded graphically in Fig. 84 (p. 265). The mouth of the estuary has been fixed, of necessity somewhat arbitrarily, at King Road off Avonmouth ; above this point extends the Severn estuary, below it the Bristol Channel. Water samples, at springs and neaps during both summer and winter, were taken at a series of points for some forty miles below and above King Road. The results of these are recorded and, below them, the total number of animals collected between tide marks by Mr. Bassindale. (These are not complete records but comparable because made by the one person.) At Porlock Weir, near the western end of the coast of Somerset, the water approaches full salinity, which is attained off the coast of North Devon. Passing up the Channel the salinity falls steadily, but to a much greater extent in winter when more fresh water is descending from the estuary. In the Severn estuary the area of brackish water moves to and fro, thrusting further up at springs, falling back at neaps, and extending highest of all in the summer months when the head of the estuary reaches almost to Gloucester.

The number of shore species decreases with the salinity of the water. At Porlock Weir 144 species of animals were collected, at Weston-super-Mare and at Portishead just over 50, still largely marine animals well adapted for the extremes of shore life. In the estuary numbers fall off still more and the animals found consisted of true brackish-water species, such as *Gammarus zaddachi*, species of *Corophium*, the brackish-water acorn-barnacle, *Balanus improvisus*, and the ragworm, *Nereis diversicolor*. In the lower reaches are some snails and the mud-dwelling bivalve, *Macoma balthica*. In the Tamar estuary this has been recorded in densities up to 5,900 per square metre and even higher in the Mersey estuary. Like the closely allied deposit-feeder, *Scrobicularia plana*, which in many estuaries extends into even more brackish waters, it finds abundant food in the rich organic content of the estuarine muds.

Among seaweeds, the green *Enteromorpha*, as would be expected, extends highest and occurs at Epney. The highest point attained by a brown alga is Purton where a stone breakwater provides attachment for *Fucus vesiculosus*. Sharpness is reached by *Pelvetia canaliculata*, *Fucus spiralis* and *Ascophyllum nodosum*. The remaining common brown sea-weed, *F. serratus*, was not encountered higher than Shepardine.

Few of these animals and none of these plants are capable of move-ment except in the early stages of life. Some shore creatures, however, move to and fro with the seasons. This is especially true of the shrimp, *Crangon vulgaris*, which is very numerous in these waters, where it is extensively fished in the fixed engines described in Chapter 19. The seasonal abundance and movements of this animal were studied by Dr. A. J. Lloyd who found that they migrate into the estuary during the summer but leave it in the winter. The males are more susceptible to low salinity than the females which are the first to return in the spring and last to leave in the winter, although during the summer females carrying eggs always migrate down the Channel to fully saline waters before the young hatch out. Temperature comes into the picture as well as salinity; the shrimps can stand low salinities in the warmth of summer but not in the cold of winter.

Then there is the question of turbidity. An excellent opportunity for observing its effects was found at Portishead where there is a large dock in the still waters of which sediment falls to the bottom although the salinity is the same as in the turbid waters outside. A comparison of the population of dock and shore made by Dr. R. D. Purchon revealed the presence of various animals, notably suspension-feeding bivalves, with polyzoans and sea-squirts which feed in the same manner, exclusively within the clear waters of the dock. There was also a thriving population of the common jellyfish, *Aurelia*, both the small hydroid stage on the pier piles and, in the summer, the pulsating medusoids, which was probably similarly barred by the high silt content from life in the open waters of this part of the Channel.

The shores of estuaries with their extensive mud banks and limited fauna are seldom ideal collecting-grounds for the naturalist. Interest resides in the effects of declining salinity, with species after species failing to make the necessary functional adjustments until only the truly brackish-water animals remain. In their success is revealed the means whereby the barrier of entry into fresh waters from the sea has been overcome. With the separation of brackish waters from contact with the sea certain species have become inhabitants of fresh water ; this has indeed been the fate of some of the former inhabitants of the Zuider Zee since, by the erection of the dyke, this former brackish-water arm of the sea has been converted into the Yssel Lake. In other cases there has been a direct invasion of the rivers from the sea and within recent times the zebra-mussel, *Dreissensia polymorpha*, and the freshwater hydroid, *Cordylophora lacustris*, have moved upstream from estuaries and become permanent members of the freshwater fauna of this and other European countries.

CHAPTER 18

DISTRIBUTION AND FLUCTUATIONS
OF THE SHORE POPULATION

"There is a river in the ocean. In the severest droughts it never fails, and in the mightiest floods it never overflows. Its banks and its bottoms are of cold water, while its current is of warm. The Gulf of Mexico is its fountain, and its mouth is in the Arctic Seas. It is the Gulf Stream. There is in the world no other such majestic flow of waters. Its current is more rapid than the Mississippi or the Amazon, and its volume more than a thousand times greater."

M. F. MAURY : *The Physical Geography of the Sea.* 1855

T H E population of British shores varies from place to place and also, to a lesser degree, from time to time. Apart from the obvious differences between the inhabitants of the varying types of shore, many instances have been cited in foregoing pages of the restriction of particular species, notably to northern and eastern or to southern and western coasts. Broadly speaking the British marine fauna may be divided into three groups—boreal (or north temperate), arctic-boreal and lusitanian (or mediterranean) species. This country lies within the north temperate or boreal zone and were it not for ocean currents all our marine animals and plants would be boreal species generally characteristic of that zone, except for a few on extreme south-western shores where some lusitanian species extend. These latter include the common octopus, *Octopus vulgaris*, the sponge-crab, *Dromia*, the spiny lobster, *Palinurus*, and also the very interesting little air-breathing sea-slug, *Onchidella celtica*, which clings like a rounded blob of black rubber on to rocks at and above high water level around the coasts of Cornwall.

The prime influence on our marine population comes from the warm waters of the North Atlantic Drift, ultimately derived from those of the Gulf Stream. These bathe our western coasts and extend for some distance up the English Channel and to a greater extent around the north of Scotland and southward into the northern North Sea. Where this current impinges directly upon southern and western shores it brings warmer waters in which the hardier of the lusitanian species can exist. Where it mixes with arctic waters in the north it tempers their coldness to produce the warmer conditions of the zone inhabited by arctic-boreal species. Some of these spread southward from the coast of Scandinavia with the water currents that flow into the North Sea and have established themselves along our north-eastern shores.

Thus it comes about that animals such as the rock-boring sea-urchin, *Paracentrotus lividus*, which is common in the Mediterranean, extend around the western shores of Ireland and further north along those of the Hebrides, while the viviparous blenny, *Zoarces viviparus*, an arctic-boreal species, occurs on our north-eastern but never on our western or southern shores. Although no absolutely hard and fast rule can be laid down, in general it may be said that animals found all around our coasts are true boreal species. The barnacle, *Balanus balanoides*, the red starfish, *Asterias rubens*, the small sea-urchin, *Psammechinus miliaris*, the larger hermit-crab, *Eupagurus bernhardus*, are all good examples. Those that occur only on the south and west, such as the other high-level barnacle, *Chthamalus stellatus*, the spiny starfish, *Marthasterias glacialis*, the urchin, *Paracentrotus*, and the smaller hermit, *Eupagurus prideauxi*, are lusitanian species, while those found exclusively on the east, for instance the small green urchin, *Strongylocentrotus dröbachiensis*, the viviparous blenny and the anemone confined to Caithness, *Phellia gausapata*, belong to the arctic-boreal fauna.

The major factor controlling the distribution of these, and indeed all, marine animals is temperature. Around Great Britain the surface temperature of the sea varies from a maximum of about 16° C. and a minimum of 8° C. in the south-west to 13° C. and 4° C. respectively off the north-east of Scotland. The temperatures on the shore fluctuate more widely because of the much greater effect of air temperatures. Temperature acts in two ways, on the individual animal and on reproduction. Marine organisms can exist only within a certain range of temperature, and a much more restricted range than that normally experienced by freshwater or land animals and plants, because the sea is a very constant medium and alters in temperature very slowly and to a limited extent. Many animals taken from the warm waters of the Mediterranean would quickly die in the winter temperatures around British coasts while northern animals would be unable to exist in the summer temperatures of Mediterranean waters. The effect of temperature on reproduction is still more important because an individual animal can exist within a much wider range of temperature than that in which it can spawn. An example will make this clear. The native British oyster, the boreal species, *Ostrea edulis*, lives and spawns around our coasts. Spawning begins when the temperature reaches about 15° C. and continues intermittently over the summer so long as the temperature remains approximately about or over this figure. Thus the oyster has a longer breeding period in the south than in the extreme north or east. The Portuguese oyster, *O. angulata*, is a lusitanian species and flourishes around the Iberian peninsula and along the Biscay coasts of France. Young individuals, or " seed," of this species are frequently purchased by British oyster-growers and relaid on English beds. There they live and grow normally but they fail to spawn because this demands an initial sea temperature of about 20° C. which is not attained in our seas. It follows that this species can never become established here. In the same way arctic animals which breed at very low temperatures, often little above the freezing point of sea water, could not become acclimatised here because the water never becomes cold enough for them to spawn. There is an upper as well as a lower limit to the temperature at which a marine animal can spawn and in arctic species this upper limit may be no higher than 4° C., to which the temperature of British seas does not normally descend.

One result of this is that the same species breeds at different times of the year in different latitudes within its range. Thus the common acorn-barnacle, *Balanus balanoides*, which is a northern species, breeds in mid-winter on the south-west of England, where it is at the extreme southern limit of its range, and later in the year in more northern and colder seas. On the other hand the lusitanian barnacle, *Chthamalus*, the northern limit of which coincides with that of Great Britain, as shown in Fig. 85 (p. 272), breeds here in midsummer. The general statement can be made that our winter-spawning animals are northern species at the southern end of their distribution, those that breed in spring or early summer are in the middle of their range, and summer spawners are southern species at the northern limit of their range. The first and last of these cannot extend respectively further to the south or to the north (unless warm or cold currents influence the temperature) because the temperature would be either too high or too low for reproduction. It is for this reason that *B. balanoides* disappears to the south, and *Chthamalus stellatus* to the north, of Great Britain.

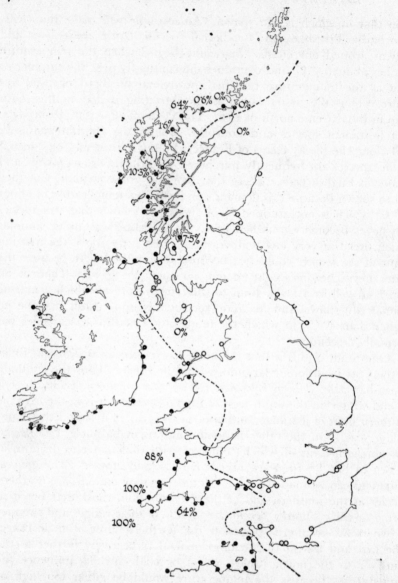

FIG. 85—Distribution of the acorn-barnacle, *Chthamalus stellatus*, around British coasts. O indicates absence, ● indicates presence. Percentage figures show the proportion of *Chthamalus* in the total population of barnacles (*Chthamalus* and *Balanus balanoides*) between tide-marks. (After Moore and Kitching, *Journal Marine Biological Association*, Vol. XXIII, p. 529.)

Plants are influenced in the same way. The eel-grass, *Zostera marina*, is interesting in this connection. Below 10° C. even growth ceases although life persists, above that temperature it starts to grow and seeds germinate, but not until it exceeds 15° C. can the flowers open and seeds be formed. Even then a sudden drop in temperature will destroy the flowers. Should the temperature rise above about 20° C., then growth again ceases and parts of the plant are destroyed. Thus the northern distribution of this plant (before the recent blight descended upon it) was controlled by the minimum breeding temperature and the southern limit by the direct effect of heat upon the plant itself.

Some animals have an exceptionally wide distribution. The common solitary sea-squirt, *Ciona intestinalis*, is a case in point. It extends from the coasts of Norway, around the shores of Great Britain, and south into the Mediterranean. But studies on the temperature at which breeding and development are possible in these different regions have revealed that, although all the animals are anatomically identical, there are several races which breed within different ranges of temperature. We may, therefore, speak here of functional or physiological, as distinct from structural, subspecies and the wide distribution of this sea-squirt is due to the presence of probably three distinct physiological subspecies. This raises the very interesting point that in the course of evolution animals may change functionally without any accompanying structural alteration. Thus the two species of European oysters differ both in shell and body characters as well as in breeding temperatures, but *Ciona* has altered only in respect of the latter. Had there been corresponding structural changes then three *obviously* different subspecies, to each of which we could have given a different name, would have come into being. Such has indeed happened in the case of *Gammarus zaddachi* noted in the preceding chapter. Some species range the world over, the best instance being the jellyfish, *Aurelia aurita*, which is found in all seas from the Arctic to the tropics. It does not appear to vary significantly in structure but clearly can breed at widely different temperatures and, although the matter has never been scientifically studied, it is almost certainly made up of many physiologically distinct subspecies breeding at different temperatures.

Apart from temperature, another factor, less easily explained, influences the distribution of marine animals. It has been found that the Atlantic water carried by the ocean drift against these islands has a planktonic population different in important respects from that of the water enclosed within the North Sea, English Channel and, to a less extent, the Irish Sea. The chemical and physical differences between these open and enclosed waters are minute but they nevertheless constitute different environments for marine life. It is possible to determine the extent to which the Atlantic water penetrates into these enclosed seas by examining the plankton. Certain members of this, in particular species of the little arrow-worm, *Sagitta*, constitute biological indicators and by following the distribution of one of these, *S. elegans*, we can determine the penetration of rich Atlantic water which fluctuates from year to year and on which the fertility of the enclosed waters is to a large measure dependent.

The same obscure property in Atlantic water which is responsible for the presence of its characteristic plankton would also appear to influence the distribution of certain shore-living animals and in particular the already much discussed barnacle, *Chthamalus*. If reference be made to the map in Fig. 85 (p. 272), in which the distribution of this species in relation to that of *B. balanoides* is shown, it will be seen that *Chthamalus* extends up the English Channel in steadily decreasing numbers as far as Swanage, up the southern entrance to the Irish Sea as far as a line drawn north-westward from the south of Anglesey, but hardly at all down through the North Channel, and that it fails to round the north-east point of Scotland and penetrate into the North Sea. While its failure to achieve the latter is probably due to low temperature, this can neither explain its inability to pass further up the English Channel nor its inability completely to colonise the shores of the Irish Sea when it flourishes far to the north of them where the temperatures are significantly lower. But its distribution does correspond in general with that of Atlantic water, and one is driven to the conclusion that the presence of this barnacle is dependent on some obscure but essential constituent of such water.

It has also been suggested that the distribution between tide marks of the large sea-urchin, *Echinus esculentus*, is influenced in the same way. This animal is very widely distributed, ranging horizontally from the north of Norway to Portugal and vertically from the shore down to over 1,200 metres. It therefore lives in temperatures of from 4° to 18° C. and so is probably made up of several physiological subspecies like *Ciona*. It has been reported living on the shore on the north coast of Spain, on the south-west of France, at Mousehole in Cornwall, along the west and north coasts of Ireland and Scotland, along stretches of the east coast of Scotland and northern England, and finally along the great extent of the Norwegian coast. This certainly approximates to the regions swept by Atlantic waters. It must be remembered that, unlike *Chthamalus*, the urchin occurs off-shore, wherever the bottom is suitable, along all European coasts north of Portugal, it is only its upward migration on to the shore that appears to be influenced by Atlantic water. The nature of the link between cause and effect is hard to imagine but marine animals are influenced by such excessively minute differences in the properties of the complex medium of sea water that the possibility of its existence cannot be disregarded.

To the effects of water currents on the enrichment of our shore fauna may be added the activities, usually unintentional, of man. The depleted oyster beds around the coasts of England have been stocked, as we have seen, by relaid Portuguese and also by an American species. The latter, *Ostrea virginica*, has much the same breeding temperature as the Portuguese oyster and has not become established here. But with it have been imported other animals which can reproduce themselves at the temperatures which prevail within our waters and so have become added to our fauna. One is the rock-borer, *Petricola*, established here about 1890, another is the slipper-limpet, *Crepidula fornicata*, and a third the American oyster-drill, *Urosalpinx cinerea*. The two last are serious pests of oysters and have transferred their attentions to our native oyster, already seriously depleted in numbers.

The case of the slipper-limpet is worth discussing. First noted here in the eighties of the last century, it now extends from the Humber round the south coast as far as Devon. About twenty years ago it appeared on the coasts of Holland, having apparently crossed the North Sea attached to floating wreckage or to seaweed, and it has since spread north along the shores of Germany and Denmark. It is certain to be encountered around the shores of East Anglia and the mouth of the Thames, where it forms masses, often inches deep, over the surface of the bottom in sheltered creeks where oysters were once abundant. Although possessing the limpet form, it is not allied to any of our common shore-limpets but is more akin to the periwinkles. The shell has an internal ledge so that, when turned over, it has some resemblance to a rounded slipper. It does not move about and has the peculiar habit of living in chains, one animal settling on the back of another until eight or nine may be so attached (Fig. 86). The under-

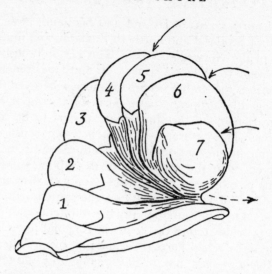

FIG. 86—Slipper-limpet, *Crepidula fornicata*, postero-lateral view of a chain of seven individuals. Animals 1-3 are female, 4 is of intermediate sex, 5-7 are males. In-going water currents (for respiration and feeding) shown by complete arrows, combined outgoing current by broken arrow. Natural size. (From Orton, *Journal Marine Biological Association*, Vol. IX, p. 444.)

most members of any chain are females, the middle ones of intermediate sex, and the youngest and uppermost, males. In such a chain there is a change in sex from male to female with advancing age, although individuals that settle direct on to a rock or oyster shell become females without any preliminary male phase. The slipper-limpet feeds like a bivalve, by means of enlarged ciliated gills, and so does not need to move about. It in no way attacks the oysters but smothers them and also competes with them for the same planktonic food.

The oyster-drill, on the other hand, is a snail closely allied to the British sting-winkle, *Ocenebra erinacea*, and has the same habit of boring into and consuming bivalves. It is a serious oyster pest in America and no less a danger here. At least one of our shore animals, the common periwinkle, *Littorina littorea*, has been transported across the Atlantic. It first appeared on the shores of Nova Scotia shortly before 1857 and is now common in that region and along the coasts of New England.

Many exotic animals are continually being introduced among the fouling organisms on the bottom of ships. Occasionally they may survive for a short time in the warm waters of docks but soon die out because they are unable to breed or to survive winter temperatures in these latitudes. Somewhat of a mystery surrounds the recent identification on British shores of a barnacle, named by Darwin *Elminius modestus*, which is common in estuaries in New Zealand and southern Australia. This has probably been carried here from the antipodes on the bottom of ships; indeed living specimens were found on a vessel which had made the passage, by way of the Panama Canal, in seventeen days. Innate hardiness combined with ability to close the shell during passage through the fresh water in the canal would explain its successful transport from the southern to the northern hemisphere. In these waters it has found temperate conditions similar to those it left in the south. Originally encountered in Chichester Harbour, it has since been found widely distributed, especially in shallow muddy estuaries, on the south-east coast between Norfolk and Cornwall. It has recently been reported from Holland. It has increased enormously in a very short time, being exceptionally hardy and prolific with a long settling period, and is already the dominant barnacle in certain areas.

Changes also occur from time to time in the distribution of our native animals. They may extend or diminish their range, and frequently wax and wane in numbers in particular areas owing to fluctuating conditions. During the latter half of the last century, the small tortoise-shell-limpet, *Patelloidea (Acmaea) testudinalis*, is said to have migrated south along the coasts of Northumberland, where incidentally it was first found by Grace Darling, then along those of Durham and finally as far south as Scarborough. All who keep constant watch on local faunas note continuous changes in the numbers of both shore-living and sublittoral animals. The sea-slugs and tectibranch snails appear from time to time in great numbers after years of scarcity. The populations of bivalves such as *Tellina tenuis* and *Cardium edule* are largely maintained by occasional good " spat falls," which may only come once in every three or four years. Apparently in some years conditions for spawning and for the subsequent survival of significant proportions of the larvae are good, whereas in other years they are not, but it is often difficult to be certain what particular factor or factors are concerned in this.

Large-scale destruction of a species does occasionally occur. This may be due to disease, the probable cause of the destruction of the

T2

eel-grass in Britain and certain cause of that of sponges around the
Bahamas. More frequently is it caused by abnormal weather, extreme
winter cold or, less frequently in this country, excessive summer heat.
These extremes of temperature are one of the major problems of life on
the shore. The major danger of winter cold is avoided by some animals
by migration into deeper and, in the winter, warmer water. This has
been noted as the regular custom among many larger crustaceans, such
as prawns, edible crabs and squat lobsters. In abnormally cold winters
they may be accompanied by shore-crabs and by shrimps which, as
we have seen, always leave the diluted waters of estuaries in winter,
and to some extent by smaller crustaceans and even periwinkles.
Extreme cold causes *L. littorea* to become inactive and it then rolls
down the shore. Shore-fishes, such as blennies, butter-fish and gobies,
also make annual retreat from the shore. Northern species may find
the summer temperature too hot; thus the viviparous blenny, *Zoarces*,
moves offshore during this season.

Many shore animals cannot migrate ; they are either attached
between tide marks or else unable, for one reason or another, to live
elsewhere. Normal winters they can survive. If they could not do so,
our shores in winter would be largely barren as they are in the Arctic
where the combination of extreme cold with ice which removes all
attached life makes existence impossible for most animals. The
majority of Arctic shore animals either migrate into deep water in
winter or else consist of annuals which, after spawning, die at the end
of their only summer on the shore. Burrowing animals can often
escape the rigours of extreme cold by movement to exceptional depths.
This has been observed in the lug-worm, the mud-burrowing amphipod,
Corophium, and to some extent in shrimps. Even then they may be
killed by lack of oxygen if the surface layers become frozen.

There are many records of widespread destruction during very cold
winters. Almost a century ago Hugh Miller, the Scottish geologist and
writer, described how the shores of the Firth of Forth were strewn with
the shells of razor-shells and other bivalves during the exceptionally
cold February of 1855. During the winter of 1904-5, when the coasts
of Lancashire were covered with ice floes for a long period, cockles
were killed in such numbers that hundreds of tons were washed up
by a succeeding gale. The exceptionally cold winter of 1928-9 was
responsible for the death of most of the sting-winkles and dog-whelks
on the oyster beds of the Essex coast although the American oyster-drill

unfortunately was little affected. At the same time the small sea-urchin, *Psammechinus miliaris*, was wiped out at Whitstable where it had previously been common, while further north, at the head of Loch Fyne, all cockles living between tide marks, but not below, suffered similar destruction. The population of *Macoma* was seriously reduced in the same region. The effects of the still more severe winter of 1946-7 were no less harmful.

The results of such destruction are often surprisingly short-lived provided there is any significant surviving population in the vicinity. The shores of Loch Fyne were recolonised by young cockles produced presumably by the survivors below low water level during 1929. With virgin shores open for re-colonisation there would be none of the normal mortality due to competition. Where a population is locally exterminated, or almost so, its place may be taken by another species which has survived. This would appear to have taken place on the shores of Kames Bay at Millport where the sand-burrowing bivalve, *Spisula subtruncata*, known locally as the aichen, which had previously been abundant, was destroyed in 1895. It became locally extinct and its place on the sandy shore was taken by *Tellina tenuis*. Within the past ten years a few living specimens of *Spisula* have been found but it still shows no sign of regaining its former great abundance in this particular area. The result of destruction of the indigenous sting-winkles on the oyster beds has been the more secure establishment of the hardier invading species from America.

Excessive heat, though rarer, has similar effects. Burrowing animals counter it (with its accompanying lack of moisture) by deeper penetration. Barnacles seem little affected ; a temperature of over 36° C. has been found within a healthy *Chthamalus*. An exceptional mortality of large cockles and of lug-worms on the sands of Morecambe Bay and in the estuary of the Dee during the summer of 1933 was probably caused by a period of exceptionally high and sudden rise of temperature during May. Dr. J. S. Huxley noted that specimens of the peacock-worm, *Sabella pavonina*, were affected by unusually high temperatures at Lough Ine, County Cork, in the same summer.

The temperate zone in which these islands lie, together with the modifying action of ocean currents, thus initially control the character of the shore population, while the activities of man have added new species from distant shores. The various populations normally fluctuate in numbers, as do those in all environments, while occasional catastrophes due to disease or abnormal weather may have far-reaching effects, often widespread if due to disease and more often local if due to weather. It is possible that such disasters on a wider scale have played an important part in evolutionary history and that we catch some glimpse of the nature of this larger drama when we observe from year to year the changes and chances that befall the inhabitants of the sea shore.

CHAPTER 19

THE ECONOMICS OF THE SHORE

"The man had sure a palate covered o'er
With brass or steel, that on the rocky shore
First broke the cozy oyster's pearly coat,
And risked the living morsel down his throat."

JOHN GAY

THE life of the shore claimed the attention of man long before the Greeks began to inquire into its nature. To primitive man it formed a valuable source of food, easily acquired and varying little with the seasons. The kitchen middens of our remote ancestors with their mounds of the shells of oysters and other molluscs remain as concrete evidence. Shells influenced man in many other ways and figured as ornaments, as symbols, as religious objects and as currency in primitive civilisations to an extent that the reader may learn in J. W. Jackson's *Shells as Evidence of the Migrations of Early Culture*.

In Britain we make comparatively little use of the resources of the shore. It is otherwise on the Atlantic coasts of France where the hors d'œuvre at many small seaside hotels may consist of an assorted collection of the local shore molluscs and crustaceans, while in the Mediterranean countries generally sea-urchins, octopuses and even large sea-squirts are also widely eaten. Even in the United States and Canada a far greater variety of molluscs is eaten than in this country. This is in part a legacy of Red Indian custom but they also possess a wider range of large edible bivalves than we do here. However, the most prized of all, the soft-shell clam, *Mya arenaria*, used in the preparation of clam chowder and in clam bakes, is equally common but entirely neglected as a source of food in this country.

A survey of the edible or otherwise commercially valuable products from the shores of the world would demand a volume to itself, but those from British shores, apart from a few fish caught by seining, are confined to bivalve and univalve molluscs, a few crustaceans, and also a variety of seaweeds the importance of which has considerably increased within recent years.

Pride of place must naturally be given to the oyster. No invertebrate

animal that lives in the sea is collected in such vast numbers or gives
employment to so many people in so many diverse countries. What
was once the bare necessity of life to a savage has been acclaimed
amongst the rarest of delicacies by the gourmets of both classic and
modern civilisations. No Roman banquet was complete without vast
quantities of oysters, of which the bloated Vitellius is said to have
consumed one thousand at a single meal. With such demands on
Mediterranean supplies it is not surprising that the Romans appreciated
the rich stocks of " natives " discovered by them at Richborough in
Kent. This flat, rounded oyster was regarded as superior in flavour to
the deeper, more irregularly shaped oyster of the Mediterranean, and
well worth the labour and expense of importing into Italy.

With the passing of Roman domination the oyster recedes somewhat from history, the subject of occasional mention in the Middle
Ages, but resumed its former importance in Tudor days, from which
period the Colchester historians begin the history of their famous
fishery. The stocks thrived despite increasing demands from the
steadily growing cities. There was a time when the native, legally
defined as an oyster, " spatted and reared in the rivers and creeks
between the South Foreland and Orford Ness," was sold for eightpence
a bushel and when Sam Weller could comment on the remarkable
connection between poverty and oysters. But during this century
stocks have fallen so low that eightpence is nearer the price of a single
oyster than of a bushel, and many of the so-called natives we now
encounter were spatted on French or Dutch beds or are inferior
Portuguese and American species, relaid on the classic beds of the
native oyster.

In the years immediately following the 1914-18 war a mysterious
malady spread through the stock of native British oysters and those of
the same species in France and Holland. Numbers sank perilously low
and have never really recovered, while the spread of slipper-limpets
and oyster-drills, and more recently of the imported barnacle, *Elminius
modestus*, have increased the problems facing the oyster grower. In
many countries, notably France, the United States and Japan, oysters
are cultivated in a variety of ways most suited to the local species
and conditions. Here we have done little beyond the spread of
clean shell, or " cultch," on which the oyster spat can settle after
planktonic life. But natural spat-falls remain poor, possibly because
the population of oysters is so low. The major hope for the future

probably lies in the further development of oyster-breeding in tanks which has been conducted with steadily increasing success by the scientific staff of the Ministry of Agriculture and Fisheries at Conway in North Wales. The problem is not easy because suitable food in the form of very minute plants has to be available for the nutrition of the larvae before they settle and the physical and chemical conditions in the water have to be maintained within narrow limits. Much also remains to be done in improving the natural beds before the native can regain its former abundance.

Although oysters may be found on the shore at low tide they are in the main inhabitants of shallow tidal creeks and never exposed. This is not true of the mussels and cockles, which rank next in importance, and are far and away higher in numbers. They are truly products of our shores. Mussels are widely collected for bait but also for human food, and are excellent when suitably cooked. They are cultivated in France and elsewhere but natural beds provide for the demand in this country.

A flourishing mussel bed with the animals attached in dense clusters and nourished by the plant plankton in the waters that flow over it, represents one of the densest accumulations of animal life (Pl. XVII, p. 138). In the words of an official publication, " It has been calculated that an acre of best mussel ground will produce annually 40,000 lb. of mussels, equivalent to 10,000 lb. of mussel meat with a ' fuel ' value of 3,000,000 calories and a money value of about £250, and this at the cost of practically no capital expenditure and only such labour as is involved in transplantation to prevent overcrowding, and to secure the best conditions for growth and fattening. No known system of cultivation of agricultural land can produce corresponding values in the form of animal food. The average yield in beef of an acre of average pasture land is reckoned to be 100 lb., equivalent to 120,000 calories and valued at, say, £7 10s. The yield of rich fattening pasture may be as high as 190 lb., equivalent to 480,000 calories and valued at, say, £14." (The values stated above, it may be noted, are those prevailing before 1939.) These figures indicate the enormous advantage gained by suspension feeders, oysters and cockles no less than mussels, to which new supplies of food are constantly being carried by tidal movements.

The major problem presented by the use of mussels as human food is the danger of contamination with domestic sewage from the towns bordering the estuaries where they often live. Bacteria are freely taken in with other suspended matter by the mussels and, although themselves unaffected, they can become a possible medium for the dissemination of pathogenic bacteria such as those of typhoid. This danger of infection from shellfish has certainly been an important factor in their neglect as human food. The ideal remedy would be to prohibit the discharge of untreated sewage near beds of shellfish. This being at present impracticable, either the sale of mussels from polluted waters has to be prevented by law or else some means found of cleansing them so that they are fit for human consumption.

Mussel purification in this country began at Conway after the sale of local mussels had been banned in 1912 owing to pollution. Concrete tanks were erected and gradually the modern methods were developed. The mussels are spread two deep on wooden grids in shallow tanks and thoroughly hosed to cleanse the outside of the shells. Sea water which has been sterilised by the addition of chloride of lime, any free chlorine being then converted into common salt by the action of sodium thiosulphate, is then admitted and the mussels remain in this water for one day. During this period the contents of the gut and of the gill cavity, in both of which bacteria may have accumulated, are discharged in strings of sticky mucus. The water is then run off and the tank refilled with sterilised water which remains for a similar period. Finally the outsides of the shells are thoroughly cleansed with chlorinated water, the only time that free chlorine is introduced, and the cleansed mussels are packed in sterilised bags which are officially sealed and then sent to the market.

This method of cleansing has proved highly successful. Additional cleansing plants have been erected in various parts of the country and recently methods have been worked out for using the same volume of sea water repeatedly so that cleansing can be carried out at inland towns. The problem of purifying oysters has also been attacked. These animals are both less liable to take in bacteria and more difficult to purify when they are polluted. It is a question of temperature. Oysters do not open their shell valves and function normally in the winter months so that the water used in purification has first to be heated to a temperature at which the oysters will open and cleanse themselves.

Cockles are less liable to pollution. Although they do occur in polluted estuaries, they are commonest on open sandy beaches. Moreover, unlike oysters and mussels, they are seldom sent alive to the market but are first boiled, or else, as at Leigh-on-Sea in the Thames estuary, steamed. This causes the bodies to come away from the shells from which they are separated by sieving. After washing in clean water they are then sent to the market without further treatment, or else salted or preserved in vinegar.

Despite the great numbers that are collected annually, over 300,000 cwt. in England and Wales, there is no apparent fall in the vast populations on the sandy shores where they thrive. The young cockles often settle in patches of soft, oozy mud and in these "nurseries" they may form a compact layer just below the surface. As they grow, the boundary of the nursery extends on to the surface of the sand around, but in the centre the surplus population cannot be disposed of in this way and many animals are forced to the surface and carried away by currents either to destruction or to stock other regions. Slower growth in the winter is reflected in the appearance of well-marked rings on the surface of the shell. From these the age can be estimated : an average marketable cockle, over an inch long, is usually three or four years old. Life is precarious. Cockles harbour a number of parasites, they are attacked by starfish, bored by snails, eaten by black-headed and other gulls when the tide is out and by flatfish which swim over the sand when the tide is in. The often devastating effects of frost and of excessive heat have already been recorded, but storms may dislodge them and throw them above high water level, currents may carry them to unfavourable areas, while streams or rivers by slight alteration of their course may undercut the beds and carry whole populations to destruction. Cockle-gatherers often destroy many more than they collect by leaving small animals at the mercy of currents or of the encircling gulls. That the cockle flourishes despite all that nature and man can do is a tribute to its virility as a species and affords further evidence of the outstanding success achieved by molluscs when they assumed the bivalve form and developed their gills as organs for collecting suspended planktonic food.

Cockles are gathered in a variety of ways, by scraping or digging, often with the aid of short-handled rakes with large teeth or with sickle-shaped scrapers. In the Morecambe Bay fishery the gatherers tread the sand with bare feet to bring the cockles to the surface, while the " jumbo," a wooden framework with a base board resting on the sand, is used for the same purpose on some Lancashire shores.

Scallops, both the smaller queen-scallop and the larger *Pecten maximus*, known as the clam in Scotland, are much prized as food, but they are caught at some depth and cannot be included among the products of the shore. Of our univalve molluscs, limpets, whelks and in particular the common periwinkle, or winkle, are collected for human consumption. Stephen Reynolds, in *The Poor Man's House*, has described the excitement of winkling. "Underneath the stone, clinging to it and lying on the bed of the pool, were so many large winkles that instead of picking them out, I found it quicker to sweep up handfuls of loose stuff and then to pick out the refuse from the winkles. When Uncle Jake came across an unusually good pocket he would call me to it and hop on somewhere else. There was an element of sport in catching the dull-looking gobbets so many together. I soon got to know the likely stones—heavy ones that wanted coaxing over—and discovered also that the winkles hide themselves in a green, rather gelatinous weed, fuzzy like kale tops, from which they can be combed with the fingers. They love, too, a shadowed pool which is tainted, but not too much, by decaying vegetable matter. Uncle Jake likes the stones turned back and then replaced ' as you finds 'em.' "

Of the edible crustaceans that may be found on the shore, crabs, occasional lobsters, prawns and shrimps, only the last are normally collected there. The others, apart from a few prawns, are fished or trapped in shallow sub-littoral waters. Off the coasts of Lancashire and Essex shrimps are netted from small vessels, but on many sandy shores they are caught in nets pushed along in shallow water or else in a simple trawl pulled by a horse and cart. Without doubt the most interesting of the shrimp fisheries are those carried out by the agency of " fixed engines " along the shores of the Bristol Channel and estuary of the Severn and which, because they may not long survive and also because of my particular interest in them, I feel I should describe. Probably from prehistoric, and certainly from Saxon times, the ebb and flow of the greatest tides in Europe over the wide mud flats has been exploited by the erection on the foreshore of basketwork traps or staked nets which are visited by the fishermen at low tide. The water as it rises over these fixed engines is strained through them, leaving behind the shrimps, the occasional prawns and the various fishes which inhabit these waters and feed upon the invertebrates in the mud banks.

These shore fisheries flourished for many centuries. They are recorded in Domesday Book and even in 1851 those along the coast of Somerset alone had an estimated annual yield of £10,000. With the rise of the deep-sea fishing industry in the latter half of the last century they suffered the common decline of all inshore fisheries and to-day only a few survive for catching salmon and shrimps, although a few stow-nets are still erected at Weston-super-Mare for catching sprats in the winter. In the Severn estuary, on the coasts of Gloucestershire, basketwork traps are employed. They are either putchers or putts, the former being used during the summer only, for trapping salmon. They are conical wicker-baskets five and a half feet long and two and a half wide at the mouth, fixed to stakes and arranged in ranks one above the other, the whole forming a putcheon weir which may be fifty yards long and ten feet high. The salmon, when attempting to swim from the shallows at the ebb, get jammed tight in these baskets which occasionally also entrap sturgeons, small sharks, the large marine lampreys and even a porpoise.

The putts, which are our more immediate concern, are also conical but larger, up to fifteen feet long and five to seven feet wide at the mouth. Each consists of a wide frontal region known as the kype, a central region or butt which contains a valve of split withies which retains any large fish, and a hinder part called the forewheel which also possesses a valve and is closed by a wooden bung. The kype is fixed between six stakes while the butt and forewheel are secured in the clefts of two Y-shaped stakes. These traps are arranged in rows facing upstream because the fishermen rely on the seaward-flowing waters of the ebb tide to bring shrimps and fish into them. The former are removed by loosening the forewheel and pulling out the bung. Descriptions and photographs of both types of these basketwork traps will be found in Mr. Brian Waters's *Severn Tide*, which gives a delightfully written account of the varied fishing activities in the Severn River and upper estuary.

On the Somerset coast baskets are replaced by nets. The largest-remaining shrimp fishery in the Bristol Channel, on the mud flats at Stolford, contains some four hundred of these nets although within living memory over one thousand were in use and in earlier times very many more. These shrimp or hose nets are cone-shaped with a rectangular mouth five and a half feet long and four high which is tied at the four corners to two stakes (Pl. XXXIIb, p. 241). The narrow end of the net, which is held open by two circular hoops of cane and contains a non-return valve of netting, is secured to a stake behind. Like the putts, the nets point upstream and fish on the ebb.

A surprising assortment of fish and invertebrates is collected within these spacious nets and baskets. The fish are in the main small and of little commercial value these days. They have included such rare animals as the sea-horse, *Hippocampus guttulatus*, allied to the pipe-fishes and a true lusitanian species common on the southern Biscay coasts of France and from there round into the Mediterranean, and the completely transparent ribbon-shaped leptocephalus larvae of the conger-eels which are hunted with dogs near by at Kilve. Although the adult fish is so common, the larvae are surprisingly rare, developing in deep water where the adults migrate for the one spawning which, as with the common eel, ends the life-history. An equally unusual invertebrate is the glass-shrimp, *Pasiphaea sivado*, which is very common in this area but otherwise largely confined to deep water. It is excessively compressed and practically transparent. The female carries a relatively small number of large yellow eggs from which the young hatch out at an advanced stage, a usual feature among deep-water animals where early life in the plankton is impossible. A few large prawns, *Leander serratus*, are caught in the nets but the vast bulk of the

FIG. 87—" Mud Horse " used in shrimp fisheries at Stolford, Bridgwater Bay. (From Harper, *The Somerset Coast*. Chapman & Hall.)
(This charming sketch demanded inclusion despite the error of showing the front of the " Horse " flat instead of bent upward. C.M.Y.)

commercial catch consists of shrimps, of which many hundredweight are taken during the summer, to be boiled immediately after capture. At Stolford where the nets are secured on soft banks of mud a mile from the shore, the fishermen use a type of intertidal sledge or " mud horse " (Fig. 87, p. 288), which they push in front of them and which serves the double purpose of preventing them from sinking deeply into the mud and of carrying back the catch.

Seaweeds of the shore and of the *Laminaria* zone have found a diversity of uses, as food for man and animals, as manure, and as sources of raw material, both inorganic and organic. At a time when their value had diminished greatly, war-time demands arose which the coming of peace has not decreased and at present it seems not unlikely that the prolific crops of brown algae which fringe our northern and north-west shores may become a national asset of considerable value.

Certain intertidal species have long been used for human consumption in these islands, notably carrageen or Irish moss (*Chondrus crispus*) in Ireland and the Hebrides, the dulse (*Rhodymenia palmata*) in Scotland, and laver (*Porphyra laciniata* or *P. vulgaris*) in England and Wales, and also in Ireland where it is called sloke. The pepper-dulse (*Laurencia pinnatifida*), which has a hot biting taste, is sometimes used as a condiment in Scotland. Laver after washing is usually boiled with a slight addition of vinegar and the gelatinous mass made into cakes which are coated with oatmeal and then fried. Dulse may be eaten raw, or cooked and used as a vegetable, or dried in the sun, rolled and then chewed like tobacco.

Carrageen moss is largely used in the preparation of jellies and for a variety of industrial uses where a gelatinous material is required. The closely allied red seaweed, *Gigartina stellata*, with which it is frequently mixed and which is often indiscriminately collected with it, acquired considerable value during the recent war as a source of agar-agar. This is an extract from various red seaweeds which dissolves in water and sets to a firm jelly. This forms an ideal medium for the growth of cultures of bacteria and has also other uses in medicine in the preparation of jellies. Until 1939 the almost exclusive source of supply was Japan, where it had been known for centuries and where over 1,000 tons were produced annually. It was manufactured largely from *Gelidium corneum*, the supplies of which were increased by cultivation on stones suitably arranged in the sea. The stoppage of these supplies led to intensive investigations in many countries for suitable alternative sources. In this country, as a result of the investigations of Dr. Sheina M. Marshall and Dr. A. P. Orr at the Marine Station at Millport, methods were worked out for the successful preparation of agar from *Gigartina*, of which large supplies exist on the coasts of Wales and especially on the west of Scotland. *Chondrus* proved another source of supply, but other weeds which yield agar are too sparsely distributed to make collection worth the trouble. The demand for agar for bacteriological purposes is relatively small but a promising industry is being developed in the preparation of table jellies and other food substances from British agar.

Brown seaweeds are a possible source of food for sheep. On North Ronaldshay, the northernmost island of Orkney, there is a breed of small short-tailed sheep that lives largely on seaweed and, according to the islanders, will die if deprived of it. The meat is dark and rich in iodine from the weed. Recent experiments in Ireland indicate that *Laminaria digitata* and *Ascophyllum nodosum* are both good food for sheep, comparable with meadow hay in feeding value, although species of *Fucus* are of little value. The fucoid and tangle-weeds have long been used for manure, either after cutting fresh from the shore or by collecting the stranded weed thrown ashore after storms. They are particularly rich in potash, although the supplies of phosphorus and of available nitrogen are less than in farmyard manure. They have the special advantage of conserving moisture, while no weeds or animal or plant pests are introduced with them as they may frequently be with farm-yard manure. On the coastal lands of Ayrshire seaweeds are extensively used on the potato fields where up to thirty tons are scattered per acre.

The use of brown seaweeds as a source of raw material for industry has a long and interesting history since its origin in the early years of the eighteenth century. After collection the weeds were dried in the sun and then burnt in shallow pits to produce a hard dark-grey mass known as kelp, about one ton of which was obtained from twenty tons of weed. Soda and potash were extracted from the kelp and during the eighteenth and the beginning of the nineteenth century this provided the main source of the soda used in the manufacture of glass and soap. The kelp industry flourished along western rocky shores as far south as the Scilly Isles, but especially in the Hebrides and Orkney. Twenty thousand tons of soda worth four hundred thousand pounds were produced each year in the Hebrides alone. In Orkney up to three thousand tons was the annual crop between 1790 and

1800, the price being then about ten pounds a ton. Yet when the industry had first been introduced into these northern islands the population had opposed it with the greatest vigour, pleading " that the suffocating smoke that issued from the kelp kilns would sicken or destroy every species of fish on the coast, or drive them into the ocean far beyond the reach of the fishermen ; blast the grass and corn on their farms ; introduce diseases of various kinds, and smite with barrenness their sheep, horses and cattle, and even their own families." But finally they were attempting, and not without success, to increase the supplies of weed by covering sandy shores with stones !

After the close of the Napoleonic wars, the importation of barilla from Spain and the development of the salt industry in Cheshire provided cheaper sources of soda, so that when James Wilson made his *Voyage round the Coasts of Scotland and the Isles* in 1841 he found the kelp industry entirely destroyed. He was not to know that it was to rise and fall again before the nineteenth century had run its course. The discovery of iodine in 1811 and later of its value in medicine was the reason for the revival. The quantities of this element are small, not more than forty pounds in a ton of weed, but for a considerable time this was the only source of what had become an important product, while the accompanying production of potash was an added source of profit. During this period there were no less than twenty manufacturers of iodine in Glasgow. Iodine continues to be obtained from this source in various parts of the world but no longer in this country which now relies on importation from Chile where iodine is obtained cheaply as a by-product of the nitre industry.

Within recent years the brown weeds have suddenly assumed new importance as a source of raw material, but this time of organic compounds and in particular of alginic acid. This acid, or its salts, provides the tensile strength and elasticity essential for the survival of these plants in the surging waters where they flourish in such profusion. The massive stalks of the larger laminarian weeds form the best commercial source of this substance. Alginic acid itself can be spun like artificial silk while its salts have many and varied uses ; they can be formed and made into light boards, used as a basis for jams, jellies and cosmetics, for finishing leather, for fireproofing and so forth. An important industry can probably be established if adequate and continuous supplies of weed are available. There are various problems to solve including a satisfactory method of harvesting a crop which grows on the roughest of bottoms in what are frequently stormy seas, while it is necessary to know the speed with which the crop will be replaced by natural growth. Fundamental research is proceeding in the marine laboratories of this country, while the numerous practical aspects of harvesting, processing and manufacture are being studied by the Scottish Seaweed Research Association which, with substantial financial assistance from the Government, is hoping to exploit the abundant crops of brown algae which once formed the basis of the kelp industry along the coasts and islands of western and northern Scotland.

APPENDIX

"All knowledge is of itself of some value. There is nothing so minute or inconsiderable, that I would not rather know it than not."

DR. SAMUEL JOHNSON

ANYTHING in the nature of an exhaustive bibliography of what has been written on the subject-matter of this book is impossible : it would require a second and no smaller volume. The origin of the figures has been given and more than fifty books have been mentioned in the course of the preceding pages, but there remain a variety of others that demand mention, while at least the chief sources of scientific papers dealing with the population and the conditions of life on British shores should be recorded. Books believed to be in print at the time of writing are indicated by an asterisk ; the remainder can be obtained second-hand in some cases, or through libraries. The dates, where these are available, are those in which the book mentioned was first published, and are not therefore necessarily those of the latest edition.

Of general works on marine biology and oceanography it is necessary only to mention a few of the more easily accessible, most of which contain references to other more exhaustive or more specialised books. For this reason reference is confined to *Science of the Sea,** second edition, edited by E. J. Allen (The Clarendon Press, Oxford, 1928), *The Ocean** by Sir J. Murray (The Home University Library, Thornton Butterworth, Ltd.), *Founders of Oceanography* by Sir William Herdman (Edward Arnold & Co., 1923), *The Seas** by F. S. Russell and C. M. Yonge (The Wayside and Woodland Series, Frederick Warne & Co., Ltd., 1928) and *British Marine Life** by C. M. Yonge (Britain in Pictures, Collins, London ; reprinted in *Nature in Britain*, ed. W. J. Turner,** 1946).

The topography of our coastal lands and shores, unfortunately excluding Scotland, has recently been thoroughly described in *The Coastline of England and Wales** by J. A. Steers (Cambridge University Press, 1946), but mention should also be made of *Tidal Lands* by A. E. Carey and F. W. Oliver (Blackie & Son, Ltd., 1918) which deals with the problems of coastal maintenance.

Books dealing specifically with the shore, apart from the early volumes of Gosse and his contemporaries, include D. P. Wilson's *Life of the Shore and Shallow Sea* (Ivor Nicholson and Watson, 1935) and *They Live in the Sea** (William Collins, 1947) ; also the revision by R. Elmhirst of M. Newbigin's *Life by the Seashore* (George Allen and Unwin, Ltd., 1931) which deals particularly with life on Scottish shores. *The Biology of the Sea-Shore** by F. W. Flattely and C. L. Walton (Sidgwick and Jackson, Ltd.), although unfortunately not brought up to date since its publication in 1922, contains much of interest about the habits and mode of life of the intertidal population. In addition there are the following somewhat older but still useful books : *Animal Life by the Sea-Shore* by G. A. and C. L. Boulenger (Country Life Library), *A Naturalist's Holiday by the Sea* by A. de C. Sowerby (George Routledge & Sons, Ltd., 1923), *An Outline of the Natural History of our Shores* by Joseph Sinel (Swan Sonnenschein & Co., Ltd., 1906), *The Sea Shore* by W. S. Furneaux (Longmans, Green & Co., 1903) with two volumes by E. Step, *By the Deep Sea* (Jarrold & Sons) and *A Naturalist's Holiday* (Thomas Nelson & Sons).

The much more recent handbook for collectors, *The Littoral Fauna of Great Britain** by N. B. Eales (Cambridge University Press, 1939) not only gives an excellent general account of the majority of animals likely to be encountered on our shores but also lists the leading works of reference on the different groups. It is indispensable to the collector who can pursue the sources of knowledge to their origins should he so desire. The intertidal algae are fully described in the authoritative *Handbook of the British Seaweeds** by L. Newton (British Museum [Natural History], 1931), while *Manx Algae** by M. Knight and M. W. Parke (University Press of Liverpool, 1931) is another most valuable modern work which contains much about the ecology and mode of life of seaweeds on the shores of the Isle of Man. There are many older books on seaweeds, often worth consulting on account of the beauty of their illustrations (in some consisting actually of dried specimens), but much of the nomenclature is out of date and needs to be checked.

Reference must also be made to the long series of monographs on British marine animals and some seaweeds (including *Manx Algae* mentioned above) which was started by the Liverpool Marine Biological Committee in 1899 and is known as the L.M.B.C. Memoirs, a name still retained although the Committee ceased to exist in 1920 when publication was transferred to the Department of Oceanography,

University of Liverpool. The thirty-fourth of these memoirs was published in 1947 and others are likely to appear. Among the many publications of the Ray Society are important works on marine animals, some of which, such as Darwin's monograph on the Cirripedia (barnacles), Alder and Hancock's Nudibranch Mollusca and Stephenson's Sea-Anemones, have been referred to in the text, while others, in particular the beautifully illustrated books of G. T. Allman on Gymnoblastic or Tubularian Hydroids and of W. C. McIntosh on British Annelids and Nemertines, are listed by Eales in *The Littoral Fauna of Great Britain*. Other books which deserve mention here because they may be of real assistance to the shore naturalist are *The Life of Crustacea* by W. T. Calman (Methuen & Co., Ltd., 1911) and *Shell Life** by E. Step (Frederick Warne & Co., Ltd.). On the economic side there are *Oyster Biology and Oyster Culture** by J. H. Orton (Edward Arnold & Co., 1937) which contains a useful bibliography, and *Lobster and Crab Fishing** by W. S. Forsyth (Adam and Charles Black, 1946), while *Marine Boring Animals** by W. T. Calman (British Museum [Nat. Hist.], Economic Series, No. 10) gives a useful short account of these highly specialised organisms.

Finally there are the very numerous original papers, largely published during the past thirty years, which have provided the main sources of information for the writing of this book. They are not by any means all British in origin ; many come from workers on temperate shores on both sides of the Atlantic and to some extent in other oceans. There may be mentioned in particular the long series of papers on the ecology of the South African Coast published by T. A. Stephenson and collaborators in *The Transactions of the Royal Society of South Africa* and the *Annals of the Natal Museum*. But confining ourselves, as we must, primarily to British scientific publications, much the most important of these in this connection is the *Journal of the Marine Biological Association of the United Kingdom*. This is published by the Plymouth Laboratory and contains the bulk of the marine researches carried out at that and the other similar laboratories in this country. A glance through the title pages of the two volumes of the first series and of the twenty-six volumes of the present, larger, series which have been published at the time these words are written reveals the wide scope and interest of the contents. Here, among much else, will be found a major part of the recent work on general shore and estuarine ecology and with it many papers on the structure, development and general biology of such important intertidal animals as acorn-barnacles, molluscs such as limpets, periwinkles, dog-whelks, top-shells with oysters, cockles and many other bivalves, crustaceans ranging from small sand-dwellers to the large decapod crabs and lobsters, together with polychaete worms of all kinds and echinoderms such as the common and burrowing urchins.

The Scottish Marine Biological Association at present only publishes an annual report although, with its increasing staff at the Millport Laboratory, it may soon start to produce its own journal. The Dove Marine Laboratory at Cullercoats, Northumberland, and the Marine Biological Laboratory at Port Erin, Isle of Man, both produce annual reports containing original papers. The publications of the Government Fisheries Laboratories at Lowestoft and Aberdeen are concerned primarily with papers on food-fishes and on the hydrography, plankton and bottom fauna of the fishing grounds in the North Sea and further afield on the continental shelf.

There remain many important papers scattered through the volumes of a variety of scientific journals such as the *Philosophical Transactions* and the *Proceedings of the Royal Society*, the *Transactions* and the *Proceedings of the Royal Society of Edinburgh*, the *Journal* and the *Proceedings of the Linnean Society of London*, the *Transactions* and the *Proceedings of the Zoological Society of London*, the *Journal of Ecology*, the *Journal of Animal Ecology*, the *Journal of Experimental Biology*, the *Quarterly Journal of Microscopical Science*, the *Proceedings of the Malacological Society*, the *Annals and Magazine of Natural History*, and many others. All this is not quite as complicated as it appears, since a good recent paper on some particular shore species or on some aspect of intertidal ecology will contain a list of references to previous work on the subject ; and by following up the clues to information so provided, the interested reader will gradually gain access to at least the major sources of knowledge on the subject of his choice.

Fig. 88—Tailpiece. From Forbes, *History of British Starfishes*

GLOSSARY AND INDEX

Figures in heavy type refer to pages opposite which illustrations will be found.